International Best Practices of Public-Private Partnership

Robert Osei-Kyei • Albert P. C. Chan

International Best Practices of Public-Private Partnership

Insights from Developed and Developing Economies

Springer

Robert Osei-Kyei
School of Built Environment
Western Sydney University
Parramatta, NSW, Australia

Albert P. C. Chan
Department of Building and Real Estate
The Hong Kong Polytechnic University
Kowloon, Hong Kong

ISBN 978-981-33-6267-3 ISBN 978-981-33-6268-0 (eBook)
https://doi.org/10.1007/978-981-33-6268-0

This Springer imprint is published by the registered company Springer Nature Singapore Pte Ltd.
The registered company address is: 152 Beach Road, #21-01/04 Gateway East, Singapore 189721, Singapore

Preface

Public-Private Partnership (PPP) is a procurement method which brings the public and private sectors together in order to deliver 'value for money' public infrastructure. The concept has grown over the years with many governments particularly from developing economies/countries seeking to enhance infrastructure growth and development through this scheme. Essentially, PPP has become a major infrastructure procurement policy adopted by many governments globally to address the rapid increase in demand for infrastructure due to the increase in population growth. However, in spite of the attempts and considerable efforts by governments, there are many reported problems and challenges faced by local and international private developers and local government authorities. Further, the PPP concept has now become a global concept, where investors, consultants, financiers, and public officials are currently engaged irrespective of their cultural backgrounds and/or geographical differences. Therefore, it has become imperative for stakeholders in this contemporary time to fully understand the practices of PPP across regions and countries with diverse economies and cultures. A thorough and in-depth knowledge of the practices of PPP from diverse cultures particularly developing and developed economies will adequately inform academics and international private investors of the emerging practices adopted by different countries in PPP.

Therefore, this book aims to provide in-depth analysis and discussion on the international best practices of PPP from developing and developed economies' perspectives. It provides strategic procedures, useful practices and information about the similarities and differences in PPP practices in developing and developed economies based on empirical evidence and case studies. Further, the book explores emerging issues and concepts in international PPP practices. This book is structured in nine chapters. The first chapter explores the basic concept of PPPs by looking at the varying definitions of the PPP concept, stakeholders involved and the process. The second chapter looks at the global development and practices of PPP particularly from developing and developed countries' perspectives. This chapter further explores the best practices adopted in past projects from these diverse regions. The third chapter looks at the motivations behind the adoption of PPP for infrastructure development in both developing and developed economies. The fourth chapter explores the critical success factors of implementing PPP projects from developed and developing economies perspectives. The fifth chapter looks at the risk assessment of PPP projects from developing and developed economies perspectives. The sixth chapter presents a model for assessing the success or performance levels of PPP projects. The seventh chapter explores the management of unsolicited PPP proposals. Specifically, this chapter focuses on the motivations for the adoption of unsolicited PPP and the effective management strategies that should be considered in their adoption. The eighth chapter explores and discusses conflict management in PPP project implementation Specifically, this chapter looks at the root causes of conflicts, the resolution mechanisms available for PPP and the prevention measures that can be adopted by practitioners. The ninth chapter provides a comprehensive best practice framework for implementing international PPP projects.

Parramatta, NSW, Australia Robert Osei-Kyei
Kowloon, Hong Kong Albert P. C. Chan

Acknowledgement

Authors would like to express our sincere gratitude to all industry experts who participated in the research studies or publications reported in this book. Further we would like to thank the Hong Kong Research Grants Council and The Hong Kong Polytechnic University for providing the scholarship funding to conduct the research studies reported in this book. Lastly, we thank Elsevier publishing Ltd, Emerald publishing Ltd, American Society for Civil Engineers and VGTU publishing Ltd for granting us the full permission to use portions of our previous publications to develop this book.

Contents

1 **Introduction to the Public-Private Partnership Concept** 1
1.1 Introduction 1
1.2 Defining Features of Public-Private Partnership 1
1.3 Models/Types of PPP Arrangements 3
1.4 The Process of Procuring PPP Projects 4
1.4.1 Project Inception Stage 4
1.4.2 Procurement Stage 6
1.4.3 Construction Stage 6
1.4.4 Operation and Transfer Stage 6
1.5 Parties Involved in PPP 7
1.5.1 Public Partner 8
1.5.2 Private Partner 8
1.5.3 Other Stakeholders 8
1.6 Chapter Summary 9
References 10

2 **Global Implementation Practices of Public-Private Partnership** 13
2.1 Introduction 13
2.2 Evolution of Public-Private Partnership 14
2.3 PPP Practice in the Developed Economies/Countries 15
2.4 PPP Practice in the Developing Economies/countries (Low- and Middle-Income Countries) 17
2.5 Experience of International Case Studies 19
2.5.1 Lekki Toll Road Concession Project (Nigeria) (Case 1) 20
2.5.2 Inkosi Albert Luthuli Central Hospital (IALCH) (South Africa) (Case 2) 22
2.5.3 Cross Harbour Tunnel (CHT) (Hong Kong) (Case 3) 24

2.5.4 Hong Kong Disneyland Theme Park (Hong Kong) (Case 4) . . . 25
2.5.5 The Ghana National Housing Project (GNHP) (Ghana) (Case 5) . . . 26
2.6 Chapter Summary . . . 31
References . . . 31

3 Reasons for Governments Adoption of Public-Private Partnership . . . 37
3.1 Introduction . . . 37
3.2 Review of Past Studies on governments' Motivations/Reasons for Adopting PPP . . . 38
3.3 Data Collection . . . 39
3.4 Differences in Governments' Reasons/Motivations for Adopting PPP in Ghana and Hong Kong . . . 41
3.5 Similarities on the Rankings of Governments' Reasons/Motivations for Adopting PPP in Ghana and Hong Kong . . . 45
3.6 Chapter Summary . . . 47
References . . . 47

4 Factors Contributing to the Success of Public-Private Partnership Projects . . . 51
4.1 Introduction . . . 51
4.2 Review of Literature on the Critical Success Factors for PPP Projects . . . 52
4.2.1 Political Stability . . . 52
4.2.2 Political/Government Support . . . 52
4.2.3 Good Governance . . . 53
4.2.4 Legal and Regulatory Framework . . . 53
4.2.5 Coordinated Leadership and Entrepreneurship Skills . . . 54
4.2.6 Sound Economic Policy . . . 54
4.2.7 Stable Macro-economic Indicators . . . 55
4.2.8 Mature and Available Financial Market . . . 55
4.2.9 Financial Support . . . 55
4.2.10 Technology Innovation . . . 56
4.2.11 Technology Transfer . . . 56
4.2.12 Strong Private Consortium . . . 57
4.2.13 Private Sector Financial Capabilities . . . 57
4.2.14 Good Project Technical Feasibility . . . 58
4.2.15 Appropriate Risk Allocation and Sharing . . . 58
4.2.16 Openness and Constant Communication . . . 58
4.2.17 Clear Goals and Objectives . . . 59
4.2.18 High Level of Commitment from Both Parties . . . 59

4.3 Differences on the Rankings of Critical Success Factors (CSFs) for PPP Projects in Ghana and Hong Kong . . . 59
4.4 Similarities on the Rankings of CSFs for Implementing PPP in Ghana and Hong Kong . . . 64
4.5 Chapter Summary . . . 67
References . . . 67

5 Risk Assessment of Public-Private Partnership Projects . . . 71
5.1 Introduction . . . 71
5.2 Review of Risk Management Studies in PPP . . . 71
5.3 Ranking of Risk Factors in PPP Projects in Ghana and Hong Kong . . . 74
5.4 Chapter Summary . . . 86
References . . . 87

6 Model for Assessing the Success Index of Public-Private Partnership Projects . . . 91
6.1 Introduction . . . 91
6.2 Review of Success Criteria for PPP Projects . . . 92
6.3 Developing a Project Success Index (PSI) for PPP Projects . . . 95
6.3.1 Selection of Critical Success Criteria for PPP Projects . . . 95
6.3.2 Identification of Critical Success Criteria Groupings (CSCGs) for PPP Projects . . . 96
6.3.3 Generating a PSI for Each CSCGs for PPP Projects . . . 98
6.3.4 Developing Overall PSI for PPP Projects . . . 101
6.4 Chapter Summary . . . 105
References . . . 105

7 Management of Unsolicited Public-Private Partnership Projects . . . 109
7.1 Introduction . . . 109
7.2 Concept of Unsolicited Public-Private Partnership . . . 110
7.3 Data Collection . . . 112
7.4 Motivations for Governments' Adoption of Unsolicited Public-Private Partnerships . . . 114
7.4.1 Comparison of Rankings Among Experts from Developing and Developed Countries on the Motivations for Adopting Unsolicited Proposals for PPP projects . . . 116
7.5 Effective Strategies for Managing Unsolicited PPP Proposals . . . 119
7.6 Commonly Used and the Most Suitable Tendering Methods for Managing Unsolicited PPP Proposals . . . 122
7.7 Chapter Summary . . . 124
References . . . 124

8 Conflict Management in Public-Private Partnership Projects . . . 127
8.1 Introduction . . . 127
8.2 Concept of Conflict Management in Public-Private Partnerships . . . 128

8.3 Data Collection 129
8.4 Root Causes of Conflict in Public-Private Partnerships. 130
8.5 Conflict Resolution Mechanisms for Public-Private Partnership Projects. 135
8.6 Conflict Prevention Strategies for Public-Private Partnership Projects. 136
8.6.1 Major Groupings of the Conflict Prevention Strategies for Public-Private Partnership Projects 139
8.7 Chapter Summary 141
References. 142

9 A Best Practice Framework for Public-Private Partnership 145
9.1 Introduction 145
9.2 Best Practice Framework for PPP Implementation (Figs. 9.1, 9.2, 9.3, and 9.4) 146
9.3 Chapter Summary 148
Reference 149

Bibliography 151

Index. 171

List of Figures

Fig. 1.1 Flow diagram of the PPP process. (Sources: (Mustafa (1999b); MOFEP (2011b)) (Osei-Kyei and Chan 2018) (*With permission from Emerald Group Publishing Ltd*)) 5
Fig. 1.2 Projects parties involved in PPP contract arrangement. (Adapted and modified from Efficiency Unit XE "Efficiency Unit" (2012b)) 7
Fig. 2.1 Private sector investments in public infrastructure development across developing regions from 1990 to 2014. (Source: Osei-Kyei and Chan (2017a) (*With permission from Emerald Group Publishing Ltd*)) 17
Fig. 5.1 Risk impact matrix for PPP projects in Ghana. (Source: Osei-Kyei and Chan (2017). *With permission from Emerald Group Publishing Ltd*) 85
Fig. 5.2 Risk impact matrix for PPP projects in Hong Kong. (Source: Osei-Kyei and Chan (2017). *With permission from Emerald Group Publishing Ltd*) 86
Fig. 9.1 Best practices for Project Inception stage of PPP process. (Osei-Kyei and Chan 2018, *with permission from Emerald*) 146
Fig. 9.2 Best practices for Procurement stage of PPP process. (Osei-Kyei and Chan 2018, *with permission from Emerald*) 147
Fig. 9.3 Best practices for Construction stage of PPP process. (Osei-Kyei and Chan 2018, *with permission from Emerald*) 148
Fig. 9.4 Best practices for Operational stage of PPP process. (Osei-Kyei and Chan 2018, *with permission from Emerald*) 149

List of Tables

Table 3.1 Background information of survey respondents from Ghana and Hong Kong ... 41
Table 3.2 Differences in the ranking of reasons for adopting PPP in Ghana and Hong Kong ... 42
Table 3.3 Quartile groupings of reasons for adopting PPP in Ghana and Hong Kong ... 45

Table 4.1 Differences in the ranking of the CSFs for PPPs in Ghana and Hong Kong ... 60
Table 4.2 Similarities in the rankings of CSFs for PPP projects in Ghana and Hong Kong ... 65

Table 5.1 Risk categories of PPP projects ... 74
Table 5.2 Summary of literature on the risk factors in PPP projects ... 75
Table 5.3 Definitions of risk factors in PPP projects ... 78
Table 5.4 Analysis of risk factors in PPP projects in Ghana and Hong Kong ... 80

Table 6.1 Success criteria for PPP projects ... 94
Table 6.2 Ranking of success criteria for PPP projects ... 96
Table 6.3 CSCGs for PPP projects and their underlying CSC ... 97
Table 6.4 Weightings for CSC and three CSCGs for PPP projects ... 99
Table 6.5 Membership functions for all CSC and CSCGs for PPP projects ... 100
Table 6.6 PSI for CSCGs for PPP projects ... 101

Table 7.1 Respondents' profile ... 112
Table 7.2 Countries of respondents ... 113
Table 7.3 Criticality ranking of governments' motivations for adopting unsolicited proposals for PPP project ... 114

Table 7.4 Results of Independent Two-Sample t-test for governments' motivations for adopting unsolicited PPP projects 117
Table 7.5 Criticality ranking of strategies for effective management of unsolicited PPP proposals 119
Table 7.6 Commonly used tendering method for managing unsolicited PPP proposals... 123

Table 8.1 Profile of respondents from China and Ghana 129
Table 8.2 Mean analysis and significant test results for the causes of conflicts in PPPs in Ghana and China 131
Table 8.3 Most suitable conflict resolution mechanism for PPPs in Ghana and China 136
Table 8.4 Mean results of conflict prevention strategies for PPPs 137

Chapter 1
Introduction to the Public-Private Partnership Concept

Abstract After the emergence of the public-private partnership (PPP) concept, several definitions and models have emerged globally. Further, different parties are involved in the delivery of PPP project. This chapter presents an introduction to the concept of PPP by looking at the various definitions and PPP models that have emerged globally. Further, the chapter explores the international process of procuring PPP projects and the key project parties involved in PPP project arrangements.

Keywords PPP Definitions · Public · Private · PPP models · PPP Process · PPP Stakeholders

1.1 Introduction

Since the emergence of the public-private partnership (PPP) concept, several definitions and models have emerged globally. Further, different parties are involved in the delivery of PPP project. This chapter presents an introduction to the concept of PPP by looking at the various definitions and PPP models that have emerged globally. Further, the chapter explores the international process of procuring PPP projects and the key project parties involved in PPP project arrangements.

1.2 Defining Features of Public-Private Partnership

Public-Private Partnership (PPP) has been defined in several ways by numerous researchers and practitioners, and each definition varies from country to country (Field and Peck 2003b). Although there is no unified definition of PPP, all definitions have common features and characteristics. Essentially, these defining features of PPP are common among almost all PPP forms or models. Examples of the varying definitions of PPP are provided below:

R. Osei-Kyei, A. P. C. Chan, *International Best Practices of Public-Private Partnership*, https://doi.org/10.1007/978-981-33-6268-0_1

The Ireland's Infrastructure and Public Private Partnership Section defines PPP as

> "*a partnership between the public and private sector for the purpose of delivering a project or service traditionally provided by the public sector*" (Department of the Environment, Heritage and Local Government, 2003b, pg. 4)

Whereas, the National Council for Public – Private Partnership of USA defines PPP as

> "*a contractual arrangement between a public sector agency and a for-profit private sector concern, whereby resources and risks are shared for the purpose of delivery of a public service or development of public infrastructure*" (Akintoye et al. 2015b, pg. 4)

In the United Kingdom, the Infrastructure and Projects Authority and HM Treasury describes PPP as;

> *a relationship formed between the private sector and public bodies often with the aim of introducing private sector resources and/or expertise in order to help provide and deliver public sector assets and services*

Also, the Government of Ghana defines PPP as;

> "*a contractual arrangement between a public entity and a private sector party with clear agreement on shared objectives for the provision of public infrastructure and services traditionally provided by the public sector*" (MOFEP 2011b; pg. 2)

Lastly, the Hong Kong Efficiency Unit describes PPP as;

> "*an arrangement where the public and private sectors both bring their complimentary skills to a project, with varying levels of involvement and responsibility, for the purpose of providing public services or projects*" (Efficiency Unit XE "Efficiency Unit" 2012b, pg. 5).

In all the definitions stated above, certain features and characteristics are common. The major features and characteristics which are common in any PPP definition are that; first, the PPP concept always involve a relationship or partnership between two parties; the public and private sectors (Carroll and Steane 2000b). "Public" in this context could be local government agencies, ministries, public departments and public authorities such as metropolitan, municipal and district assemblies, federals and provincials' authorities. "Private" in this context refers to a consortium or any other private investor or developer (Pallister and Law 2006b). Second, in any PPP arrangement, each actor or party in the partnership arrangement is considered a principal. This implies that each party (i.e. public or private sector) bargains on its own behalf, thus they do not refer to an authority elsewhere (Akintoye et al. 2003b). Third, in PPP arrangements, there is a continual or long-term relationship between the project parties. This is because PPP contracts involve a concession agreement usually spanning over 10 years (Middleton 2000). The fourth defining feature of PPP is that each participant or party brings on some resources to the partnership (Collin 1998). The resources could be material (land or money) or immaterial (technology or any symbolic values) (Akintoye et al. 2003b). Lastly, in PPP arrangements, the responsibilities and risks are shared equitably among project participants. The public partner usually acts as the facilitator of the partnership by providing land, tax exemptions and other public utilities, whereas the private partner takes the responsibility of designing, financing, constructing and operating the facility. In fact, the last feature of PPP has been emphasized by many governments in their policy guidelines.

1.3 Models/Types of PPP Arrangements

Several PPP forms have emerged in the last three decades. Most of these PPP models have very similar contractual arrangements and conditions among project parties. Essentially, different countries have customized the existing models to suit their context. More also, the type of PPP model adopted in a country largely depends on the host countries' objectives and expected level of private sector participation (Li 2003b). The most common PPP models adopted globally are described below.

Build, Operate and Transfer (BOT) and Build Own Operate and Transfer (BOOT) are the early forms of PPP contract arrangements. They were practically used in the seventeenth century for road construction and water projects. The BOT model requires the private developer to build and operate a public facility for a fixed period normally 25 to 30 years and thereafter transfer it to the government. This PPP model was widely used in the early 1960s in Hong Kong for many transport economic infrastructure projects including tunnel projects. The BOOT contract arrangement was used predominantly in Australia in the early 2000s (Arndt 2000b). This model is very similar to the BOT, however in the BOOT model, the private investor owns the public facility throughout the concession period. This implies that the insurance risk of the facility is usually borne by the private developer, whereas in the BOT arrangement the insurance risk is taken by the public partner.

Design, Build, Finance and Operate (DBFO) requires the private partner to finance, develop and make available the public facility for services over a period of time. The U.K's Private Finance Initiative, which was introduced in the early 1990s is a form of the DBFO arrangement (Broadbent and Laughlin 2003b). In the DBFO model, the public partner makes annual payments to the private consortium for the usage of the facility by the general public. In situations where the private consortium fails to make the public facility available to users, the period of interruption or unavailability is quantified and deducted from the annual payment. In this PPP contract arrangement, the public partner often extends the concession period to the private partner anytime is due.

Operation and Maintenance (OM) arrangement requires the private investor to operate and maintain an existing state-owned facility. This PPP contract arrangement is largely used in China for many water and waste facilities (Cheung 2008).

Build, Transfer and Operate (BTO) contract requires the private consortium to transfer the legal title of the public facility to the public sector after passing the completion tests. However, the private consortium will continue to operate the facility for a concession period. This PPP contract arrangement often relieves the private investor of furnishing high insurance cost during operation of the public facility (Levy 1996b). Most primary facility projects in South Korea were developed using

BTO models. Moreover, this mechanism best suit government who want to own the public facility immediately after the project completion.

Build, Own and Operate (BOO) arrangement is one of Australia's concession contracts which later in 2000 was modified. This model requires the private consortium to design, build and own the public facility indefinitely. Most government enters into such PPP contract when they feel that it would be more expensive for the public sector to operate the public facility. Stein (1994b) argued that this type of concession contract is most suitable for countries without any ideological problems with private investors owning and operating a public facility or infrastructure indefinitely.

1.4 The Process of Procuring PPP Projects

The process of procuring PPP projects differ from country to country and it largely depends on the host country's regulation and policy. Notwithstanding, almost all PPP projects will follow similar procurement process. Mustafa (1999b) identified five phases at which every PPP project will pass through. This process includes the Project inception Stage, Procurement Stage, Construction Stage, Operation and Transfer Stage. In each of the project phases both private and public have a significant role to play. A generic flow diagram of the process of procuring PPP projects is illustrated in Fig. 1.1.

1.4.1 Project Inception Stage

The process of procuring a PPP project starts from the inception stage. The major activity involved at this stage is the identification of public project by the contracting authority. Often public projects are identified from the National Infrastructure Plan (NIP), National Public Investment Plan (NPIP), National Development Plan (NDP) and any other development plan of the contracting authority. In some situations, public projects are identified through the request of proposals from individuals and private developers. However, these unsolicited proposals must be a proposed project not identifiable in any of the development plan issued by the government. Pre-feasibility and full feasibility studies are also undertaken at this stage. The contracting authority engages a private consultant who is referred to as Transaction Advisor (TA) to undertake the feasibility studies of the project. The feasibility studies examine the technical, financial, environmental, social aspects of the proposed project. Moreover, the cost of the feasibility studies undertaken at this stage is initially borne by the government and later transferred to the successful bidder. In case of unsolicited proposal, the cost of feasibility studies is borne by the private developer.

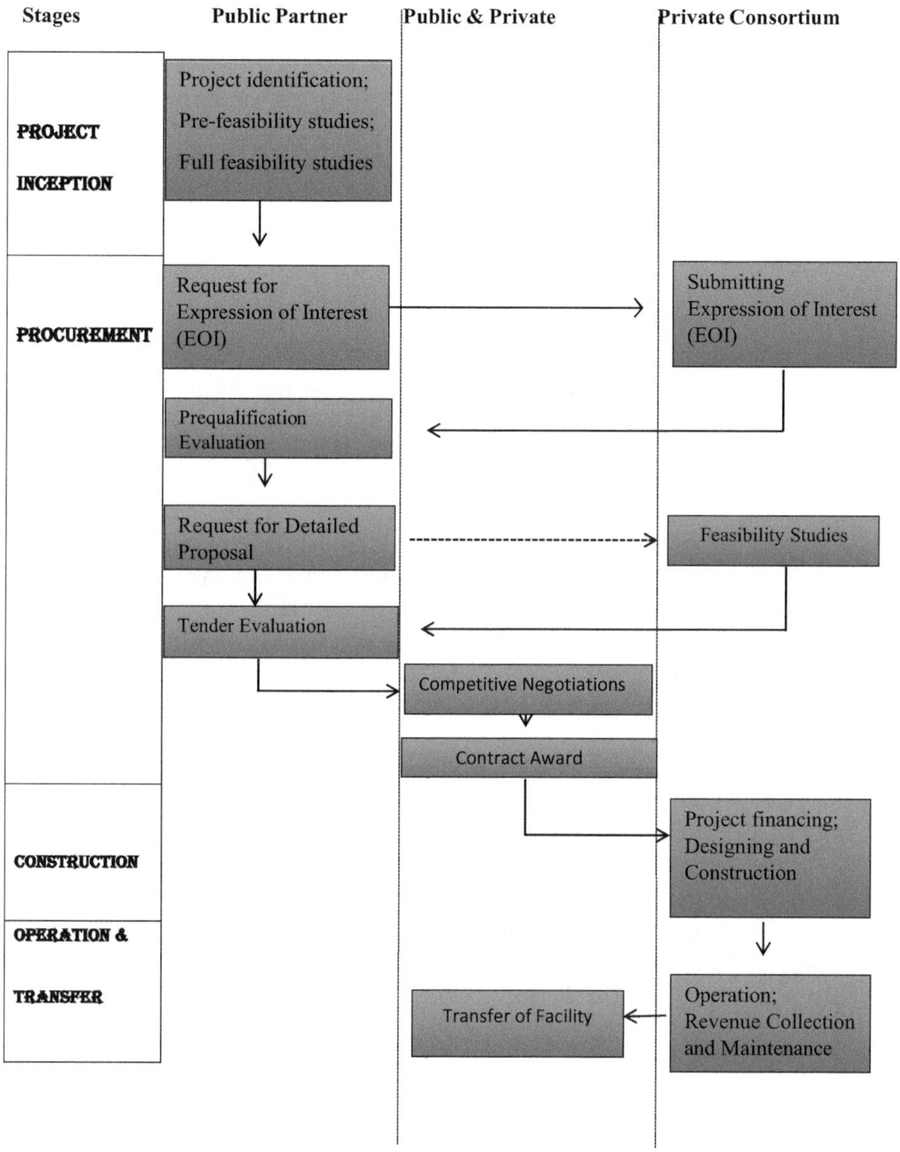

Fig. 1.1 Flow diagram of the PPP process. (Sources: (Mustafa (1999b); MOFEP (2011b)) (Osei-Kyei and Chan 2018) (*With permission from Emerald Group Publishing Ltd*))

1.4.2 Procurement Stage

PPP process varies among jurisdiction; hence this stage could be referred to as implementation stage in other countries (Li 2003b). This stage involves the request for expression of interest, evaluation of tenders, negotiations of award and financial close. After the approval of feasibility studies, a request for expression of interest is issued through suitable medium including newspapers, government websites etc. In Europe, expression of interests are advertised in the Official Journal of the European Union (Commission of the European Communities 2000b). Normally, the expression of interest is used as pre-qualification. This allows the public entity to eliminate any low standing bidder hence, facilitating the shortlisting of prospective bidders for the proposed project. Following the expression of interest; a more detailed technical and financial proposal is requested from shortlisted bidders. Upon the selection of the best bidder, negotiations are done to allow parties have a mutual agreement on the objectives and risks of the project. In the extreme case where negotiations fail, the second highest bidder is selected for negotiations. The financial close follows the negotiation process; it is after the financial close that the private partner can begin construction. In many countries such as the U.K. and Australia, losing bidders are not compensated for their commitment but other governments may compensate losing bidders to encourage more private sector commitment towards PPP investments.

1.4.3 Construction Stage

At this stage, the private partner is required to undertake the construction of the project according to the agreed output specification and quality. Very often, the private partner is compelled in the partnership agreement to employ more local content and expertise during construction. The public entity provides annual reports on the progress of the construction. This stage also involves the commission of the completed project for operation.

1.4.4 Operation and Transfer Stage

The operation and transfer stage include the day-to-day running and management of the public facility. During operation, the consortium is expected to charge the user fees agreed during negotiations, that is if direct user fee payment was agreed. In some jurisdictions, particularly the developed countries, availability-based payment is often adopted. This is where the investor is paid an annual fee based on the service quality and availability of the facility. In many countries, the concession period granted to an investor depends on the type of PPP project and infrastructure sector.

Nonetheless, averagely, concessions are granted for 25 years for road projects and much lesser years for social infrastructure such as prisons and hospitals. Also, at this stage, the public sector organizations are required to monitor the performance of the consortium. Specifically, they are supposed to ensure that the output requirements and service standards are met. At the end of the concession period, the facility is transferred to the contracting authority at no cost (Coulson 2005b).

1.5 Parties Involved in PPP

PPP as a contract arrangement brings together many stakeholders with different agendas and goals. Figure 1.2 shows an overview of the project parties involved in PPP contract arrangement. The key parties to a PPP project are the public sector and

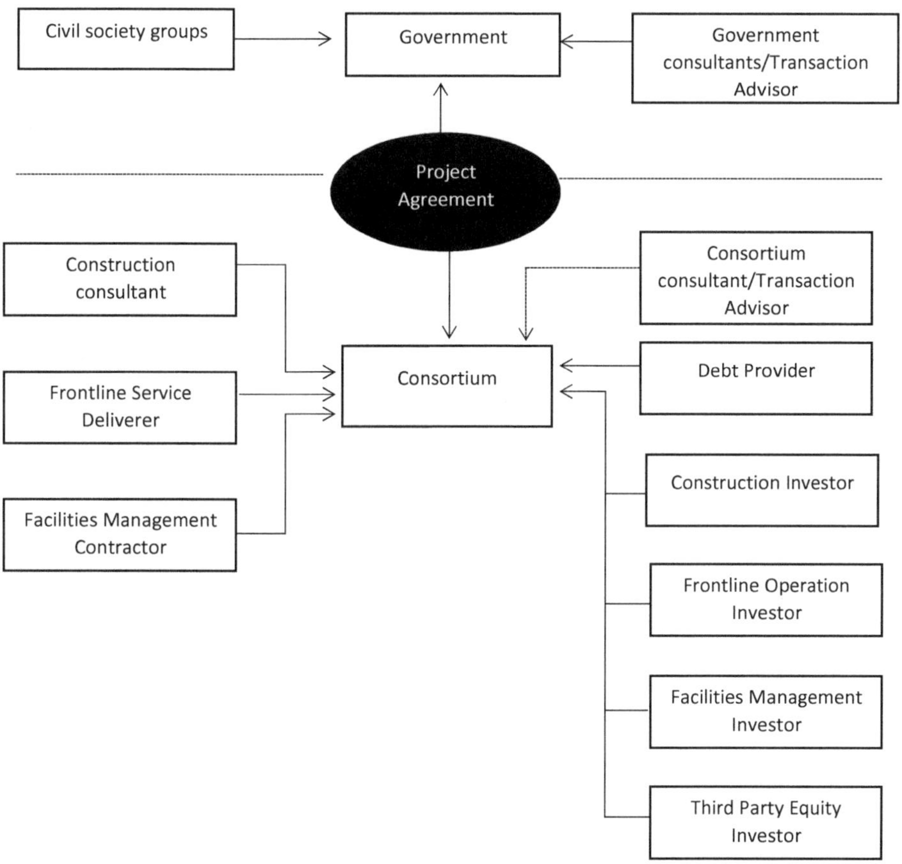

Fig. 1.2 Projects parties involved in PPP contract arrangement. (Adapted and modified from Efficiency Unit XE "Efficiency Unit" (2012b))

private consortium (Hodge and Greve 2005). Other third parties include advisors (Legal, Technical and Financial), financiers (Equity, Debt and Organisation Aids) and civil society groups. Subsequent sections discuss the key roles of the project parties in PPP contracts.

1.5.1 Public Partner

The public sector involves all government agencies and departments. However, it should be mentioned that the public sector bodies allowed to enter into PPP contracts may differ from country to country. Often the managing director or the designated minister is the one who signs the contract agreement on behalf of the public entity. Grimsey and Lewis (2004b) pointed out ten major roles that are played by the public partner/government agency. These are concession granter, network planner, funder, customer, project manager, inspector, PPP overseer, contract manager, protector of the environment and representative of the public.

1.5.2 Private Partner

The private partner is the legal entity that tenders for, finance, develops and operates the public facility. It consists of private firms and developers. These firms and developers come together to form a Special Purpose Vehicle (Estache et al. 2000). A typical SPV will include firms and developers such as engineering and construction firm, operation and maintenance company and legal firm. Private consortium usually generates funds for PPP projects through equity and debt.

1.5.3 Other Stakeholders

1.5.3.1 Financiers

The private partner will have to secure funds for the project at the financial close stage. Project funds are normally generated through equity and debt. Equity funds comes from owners or partners of the project company. Equity partners/investors involve in some portion of the PPP project development such as construction or facilities management (Hodge and Greve 2005). In fact, equity funds are a small portion of PPP capital structure. It usually constitutes about 1 – 2% of the project's capital cost. Debt financing is provided by financial institutions. In some PPP arrangements, debt financiers are part of the SPV or the project company; therefore, they tend to show great concern on the progress of the project delivery. Debt financing constitute a higher proportion of the capital structure compared to equity financing. It can be as high as 80–90% of the project's capital cost (Yescombe 2011b).

1.5.3.2 Transaction Advisors

Considering the complexities and risks of PPP projects, it is necessary for both parties to engage a consultant with adequate technical, legal and financial expertise to provide advisory services. These specialists are often referred to as "transaction advisors". Essentially, before the public sector advertises or initiate a PPP project, it engages a transaction advisor to conduct detailed feasibility studies. Also, the transaction advisor assists the public entity in the selection of a suitable private partner for the proposed project. The transaction advisor's work ends after the contract agreement is signed between project parties. Similarly, the private partner also engages a transaction advisor to assist with the negotiation of the contract terms with the public entity.

1.5.3.3 Civil Society Groups

One of the key external stakeholders of PPP arrangement is the civil society groups. Civil society groups can include trade unions, consumer protection agencies, faith-based organisations, professional associations and development NGOs. These groups ensure that the needs and expectations of the general community are considered in the development of PPP projects. They help sensitize the community and general public on the importance of developing public facilities through the PPP approach. Further, civil society groups advocate for better services and reasonable user fees for PPP projects/facilities. In some jurisdictions such as Hong Kong and Australia, the governments engage civil society groups in the identification of public facilities to be delivered under the PPP scheme. Some governments such as Ghana involve civil groups during the feasibility stages, in order to get their opinions and suggestions. It should be mentioned that in some African countries, local chiefs play the roles which are done by civil groups in the developed countries. Local chiefs are highly respected local leaders in the community; therefore, they have significant influence on the sensitization of the local community. They are usually the first point of contact for many public agencies when PPP projects are to be developed within a local community.

1.6 Chapter Summary

This chapter has explored the background of the PPP concept. It reviewed the varying definitions that has been adopted by different governments. Despite the varying definitions of PPP, similar features are realised, and these features have been discussed. The various modalities of PPP contracts have been discussed. However, it should be mentioned that the type of PPP model adopted by a government largely depends on the objectives and public sector management skills. The process of procuring PPP projects have been reviewed. The process involves 4 stages, these include; project inception, procurement, construction, operation and transfer stage. The parties of PPP projects have been discussed. This chapter has provided the general background of the PPP concept and has also provided an informative foundation for the following chapters in this book.

References

Akintoye, A., Hardcastle, C., Beck, M., Chinyio, E., & Asenova, D. (2003b). Achieving best value in private finance initiative project procurement. *Construction Management and Economics, 21*(5), 461–470.

Akintoye, A., Beck, M., & Kumaraswamy, M. (Eds.). (2015b). *Public private partnerships: a global review*. Routledge.

Arndt, R.H. (2000b). *Getting a fair deal: Efficient risk allocation in the private provision of infrastructure*. PhD thesis, Department of Civil and Environmental Engineering, the University of Melbourne.

Broadbent, J., & Laughlin, R. (2003b). Public private partnerships: an introduction. *Accounting, Auditing & Accountability Journal, 16*(3), 332–341.

Carroll, P., & Steane, P. (2000b). Public-private partnerships: sectoral perspectives. *Routledge Advances in Management and Business Studies, 19*, 36–56.

Cheung, E. (2008). Developing a best practice framework for implementing public private partnerships (PPP) in Hong Kong. PhD Thesis. Queensland University of Technology, Australia.

Collin, S. (1998). In the twilight zone: a survey of public-private partnerships in Sweden. *Public Productivity & Management Review, 21*(3), 272–283. San Francisco: Sage.

Commission of the European Communities. (2000b). *Proposal for a Directive of European Parliament and The Council on coordination of procedures for the award of public supply contracts, public service contracts.*

Coulson, A. (2005b). A plague on all your partnerships: theory and practice in regeneration. *International Journal of Public Sector Management, 18*(2), 151–163.

Department of the Environment, Heritage and Local Government. (2003b). *Policy framework for public private partnership (PPP) in Ireland*. Ireland: Irish Government.

Efficiency Unit. (2012b). *Serving the community by using the private sector – An introductory guide to Public Private Partnerships*. Hong Kong: The Hong Kong Special Administrative Region Government.

Estache, A., Romero, M., & Strong, J. (2000). The Long and Winding Path to Private Financing and Regulation of Toll Roads. World Bank, working paper.

Field, J. E., & Peck, E. (2003b). Public–private partnerships in healthcare: the managers' perspective. *Health & Social Care in the Community, 11*(6), 494–501.

Grimsey, D., & Lewis, M. K. (2004b). *Public and private partnerships. The worldwide revolution in infrastructure provision and project finance* (p. 01060-3815). Northampton: Edward Elgar Publishers.

Hodge, G. A., & Greve, C. (2005). *The challenge of public-private partnerships: Learning from international experience*. U.K: Edward Elgar Publishing.

Levy, S. M. (1996b). *Build, operate, transfer: paving the way for tomorrow's infrastructure*. New York: Wiley.

Li, B. (2003b). *Risk management of construction public private partnership projects*. Ph.D. thesis, Glasgow Caledonian University, UK.

Middleton, N. (2000). Public private partnerships — a natural successor to privatisations? http://www.pwcglobal.com/uk/eng/about/svcs/pfp/ppp.html. Accessed on Oct 2013.

Ministry of Finance and Economic Planning. (2011b). *National policy on public – private partnerships*. Ghana.

Mustafa, A. (1999b). Public-private partnership: an alternative institutional model for implementing the private financial initiative in the provision of transport infrastructure. *The Journal of Project Finance, Summer*, 64–79.

Osei-Kyei, R., & Chan, A. P. (2018). A best practice framework for public-private partnership implementation for construction projects in developing countries. *Benchmarking: An International Journal*.

Pallister, J., & Law, J. (2006b). *A dictionary of business and management*. Oxford: Oxford University Press.

Stein, S. W. (1994b). Build-Operate-Transfer (BOT): a re-evaluation. *The International Construction Law Review, 11*(2), 101–113.

Yescombe, E. R. (2011b). *Public-private partnerships principles of policy and finance*. Elsevier.

Chapter 2
Global Implementation Practices of Public-Private Partnership

Abstract Since the inception of the PPP concept, an increasing number of governments globally has shown strong enthusiasm towards the implementation of the policy. However, not all countries are on the same level of implementation, while some countries have seen significant boost in their PPP markets, others are yet to establish the enabling framework required for the implementation of the PPP policy. This chapter explores the evolution of PPP and its global implementation practices from developed and developing countries' perspective. Further, the chapter discusses the experience of high-profiled international cases to explore the failure and success factors.

Keywords Global implementation · Governments · Developing economies · Developed economies · PPP international case studies

2.1 Introduction

Since the inception of the PPP concept, an increasing number of governments globally has shown strong enthusiasm towards the implementation of the policy. However, not all countries are on the same level of implementation, while some countries have seen significant boost in their PPP markets, others are yet to establish the enabling framework required for the implementation of the PPP policy. This chapter explores the evolution of PPP and its global implementation practices from developed and developing countries' perspective. Further, the chapter discusses the experience of high-profiled international cases to explore the failure and success factors.

R. Osei-Kyei, A. P. C. Chan, *International Best Practices of Public-Private Partnership*, https://doi.org/10.1007/978-981-33-6268-0_2

2.2 Evolution of Public-Private Partnership

The private sector participation in public infrastructure delivery dates back in the early 200BCs, where the Roman government contracted some public projects to private investors (Shimron 1979). Nevertheless, some academic writers have asserted that the concept of PPP originated from the concession systems practiced between the 17th and 20th centuries mostly in France, U.K and North America (Yescombe 2011b). In the early concessions granted, private investors were allowed to charge direct fees from users to cover investments costs (Tam 1999b). Moreover, most concession contracts were awarded for public facilities such as railways, waste – water treatment networks, bridges and roads (Yescombe 2011b). Some of early concessions contracts awarded included the Brooklyn Bridge in New York built in the late nineteenth century, the water supply project awarded by James I in 1602, the Lancaster Turnpike contract awarded in 1793 and Erie Canal also completed in 1825 (Corrigan et al. 2005b; Li 2003b).

The granting of concessions and other form of private sector involvement in public infrastructure delivery continued throughout the late nineteenth century to meet the growth of industries mostly in the Europe (Beauregard 1998b; Stromback 1993b). However, after the 1929 world economic crises, many governments begun employing policies which required the establishment of more States' industries and enterprises. This actually continued after the World War II in 1939 until the 1970s (Li 2003b). Within those periods there was some declination in the granting of concessions to private investors, it was until 1980s that the programme of privatization emerged again, where most states owned assets and organizations were sold off to private investors mainly through the stock exchange market (Owen and Merna 1997b). This approach was a way to free up the economy where market forces would rather dictate the economy and to minimize public overspending. Essentially, this continued and led to the "rediscovery of concessions" with some of the initiated projects been the Channel Tunnel Project constructed in 1987 (Winch 2000b), The Dartford – Thurrock Crossing (Legislative Council Secretariat 2006b), The Sydney Harbour Tunnel (Levy 1996b) and other tunnel/ road projects initiated in some countries (Kumaraswamy and Zhang 2001b).

Later in 1992, the British government introduced Private Finance Initiative (PFI) model; a development of the old concession structure. However, this concession model adopted a shadow tolling mechanism where public authorities paid the concessionaire base on usage by motorists (Yescombe 2011b). Since 1992, the U.K's PFI model has received great attention globally with other countries developing on the model into other forms of PPP contracts (World Bank XE "World Bank" 2015d). By 2010, many countries have now adopted the PPP concept following the successful cases recorded in the U.K. and other jurisdictions like Hong Kong and Australia.

2.3 PPP Practice in the Developed Economies/Countries

The level of PPP development across countries and regions varies and this depend on the local geography, maturity of financial market, political climate and the ability to form partnerships (Egger and Startup 2006b). In this regard, countries with favourable PPP climate especially the developed countries have a more matured PPP policy as compared to developing countries.

Generally, the U.K and Australia have the well matured PPP market among the developed countries. In the U.K, since 1992, the PFI (Private Finance Initiative) model has been extensively implemented in diverse infrastructure sectors (HM Treasury and Infrastructure XE "Infrastructure" U.K 2013b). The Partnerships U.K (PUK) which succeeded the Treasury Taskforce in 2000 has contributed significantly to the current stage of U.K's PPP market. However, in 2011, the PUK was dissolved and a new organization called Infrastructure U.K. was established. The purpose of Infrastructure U.K. was to work alongside with private investors in the delivery and operation of PPP projects. In 2016, Infrastructure U.K. was merged with the Major Projects Authority to form a new body called the Infrastructure and Projects Authority. The core function of this new body is to provide expertise and advisory services to the U.K. government on the procurement and delivery of infrastructure and major projects including PPPs. As of 2018, over 700 PFI/PPP projects, which worth over £56 billion have been implemented in the U.K. in different infrastructure sectors .

Despite how well developed the U.K.'s PFI/PPP market is, there are many setbacks that are currently affecting the development and implementation of the PPP concept. Issues relating to value for money, appropriate risk allocation, excessive profit gained by the private sector and transparency have generated a lot of debate on the appropriateness of the PPP concept for developing public infrastructure in the U.K. Because of this, in 2018, the U.K. government announced their decision to stop procuring public infrastructure through the PPP scheme but will continue to operate the existing PFI/PPP projects. In fact, other political leaders have demonstrated their opposition towards the use of PPP for public infrastructure delivery in the U.K. Some have indicated that they will stop the use of PPP and return all assets and state-provided services into public ownership when they assume office. These actions and assertions have created a lot of uncertainties in the future development of the U.K's PFI/PPP market. Nevertheless, there is no doubt that over the past three decades of PPP implementation in the U.K., better and efficient public infrastructure facilities such as schools, hospitals, tunnels and roads have been provided by the private sector and that there is the need for more consultation to be done in restructuring PPP contracts so that value for money can be offered.

Australia's PPP practice is quite similar to the structure of the PFI model used in U.K. The shift from the old Build Own Operate Transfer (BOOT) model to U.K's PFI model occurred in the early 2000s (English 2006b). In Australia, the practice of PPP varies among Federal States with the Victoria State leading the implementation of this policy (Chan and Cheung 2014b). In Victoria State, the Partnership Victoria was established to offer technical and advisory services to the State government. As of 2003, approximately 10% of PPP investment has been procured with the assistance of the Partnership Victoria (OECD 2008b). In addition, as of 2017, 29 PPP projects which worth over AU$15.6 billion have been contracted out through the Partnership Victoria (State Government XE "Governments" of Victoria 2015b). These PPP projects span across different infrastructure sectors including roads, health, education and security. Other States including the New South Wales and South Australia have also established PPP units similar to Partnership Victoria (OECD 2010b). In 2008, the Australian government established an independent federal government agency called the Infrastructure Australia (OECD 2010b). The newly instated statutory body which was established under the Infrastructure Australia Act 2008 aims to provide research and advice to governments, industry and the community on the infrastructure investments and reforms in the country. Further, Infrastructure Australia focuses on setting up policies and guidelines as well as coordinating infrastructure development across the Federal States in the country (Appleby et al. 2012).

Other developed countries such as the U.S.A and Canada in the North American region have a well-structured and developed PPP market. In Canada, the Canadian Council for Public-Private Partnerships was established in 1993. It is the central PPP organization in Canada and their responsibilities are to coordinate and promote the PPP policy at the national level (Jooste 2010b). Other provinces including British Columbia, Ontario and Quebec have their well-established PPP units that offer advisory services to public departments and agencies (OECD 2010b). Since the mid-2000s, over 206 PPP projects which worth more than $63 billion have been initiated in Canada; 121 are in operational or under construction across the six Canadian provinces (CCPPP 2013b).

The U.S. PPP practice is not different from the Canadian PPP structure. Similar to the PPP organisation in Canada, the U.S has a national PPP unit; The National Council for Public-Private Partnerships (NCPPP). NCPPP succeeded the Privatization Council established in 1985 (Osborne 2002b). The NCPPP has over the years been advocating and promoting the PPP concept at the Federal, State and Local levels (Bovis 2013b). Many other States have their legislations which guide the procurement of PPP projects (Gifford 2013b). Importantly, most of the PPP projects initiated in the U.S are in the transport infrastructure sector (Wall and Wall 2013b). Essentially, between 1989 and 2011 (22 years) over 96 PPP projects which worth $54.3 billion have been initiated in the U.S transport infrastructure sector (Reinhardt 2011b). Unlike the U.K where DBFO concession type has largely been used for transport projects, the U.S employs different concession contracts including design and build (DB), design build and finance (DBF) and design build operate and maintain (DBOM) (Reinhardt 2011b).

Given the success and significant contribution of the PFI concept to the U.K, Australia, Canada and the U.S economy, other governments' in the developed countries have been encouraged to adopt the PPP concept but structuring it to suit their local context. Countries like Japan, Belgium, Italy, Spain, South Korea, Portugal and Hungary have also shown interest in the PPP concept. These countries have enacted policies and legislations which regulate the practice of this policy (Akintoye et al. 2008b; Yescombe 2011b).

2.4 PPP Practice in the Developing Economies/Countries (Low- and Middle-Income Countries)

In some years now, governments in developing countries have also demonstrated some commitment in leveraging the private sector's capital and expertise in developing public infrastructure. In this regard, the number of PPP projects over the years has increased among developing regions (Bovis 2013b). It is quite understandable because most governments in developing countries are faced with huge infrastructure investments gap of which budgetary allocations alone cannot help bridge the investment gap (Trebilcock and Rosenstock 2013b). Figure 2.1 shows private sector investments in public infrastructure development across developing regions since 1990 (Osei-Kyei and Chan 2017).

As seen in the figure, PPP investments have risen over the years particularly in developing countries in the Latin America and Caribbean (LAC) region. In LAC, Brazil has over the years contributed substantially to the increase of PPP projects. The rapid increase of projects in Brazil was realised in 2003 with 45% share of LAC PPP investments and 78% share in 2012 (World Bank XE "World Bank" 2013d). In

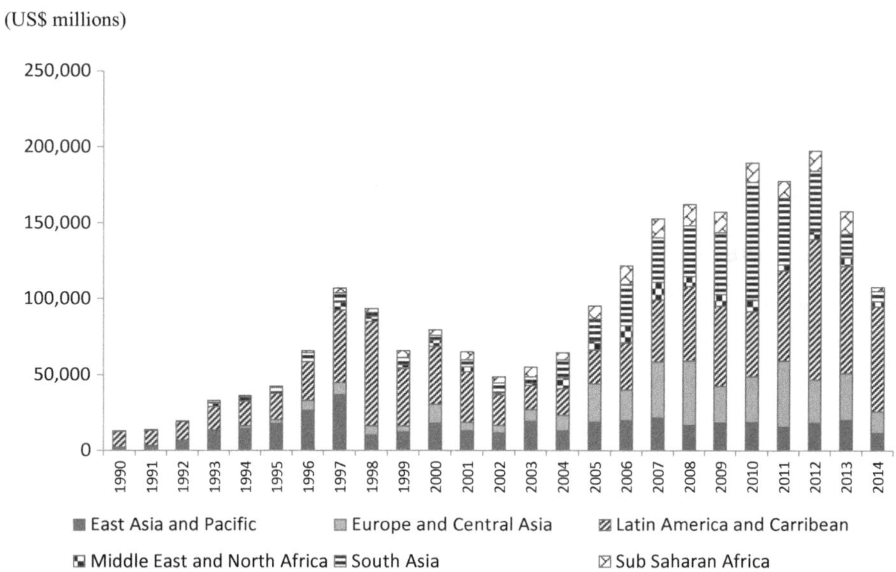

Fig. 2.1 Private sector investments in public infrastructure development across developing regions from 1990 to 2014. (Source: Osei-Kyei and Chan (2017)(*With permission from Emerald Group Publishing Ltd*))

actual fact, the Brazilian government has since 2003 shown great commitment towards private sector participation by enacting a national legislation in 2004. Other countries including Mexico and Argentina are also progressively embracing the PPP concept and they have contributed quite a number of projects to PPP investments in LAC. In 2012, a federal PPP law was enacted in Mexico, this law is therefore expected to drive the Mexican PPP market to enable the country catch up with other countries in the region.

In the East Asia and Pacific (EAP) region, improvement has been realised in their PPP market over the past two decades. This makes the region the second highest with PPP investments among developing regions (World Bank XE "World Bank" 2015d). However, China is the country with most of the PPP projects initiated in the region followed by Malaysia and Philippines. Essentially since 1990, an accumulated share of 33% of the region's private investments in public infrastructure has been contributed by China, with Malaysia and Philippines contributing about 17% and 16% respectively (World Bank XE "World Bank" 2013e).

In previous years, the socialist planned economic system in China actually limited the number of PPP projects developed in the country, however since the early 2000s, China's economic system has been opened up with new policies. This has allowed more private sector participation in public infrastructure delivery (Ho 2006b). Many Chinese local authorities are now engaging in more PPP projects than previously particularly for water projects (Meng et al. 2011b). Similarly, the Malaysian government has established a PPP unit; the Privatization and Private Finance Initiative Unit – PFI (PwC 2012b). It is expected that the PPP unit will help augment and facilitate the introduction of more public project through PPP schemes in the country to boost private investments within the region.

In South Asia (SA) region; India has been the leading country. In 2014, out of the 40 PPP projects reaching financial close, India received 36 representing 90% share (World Bank XE "World Bank" 2014). Aside India, other developing nations including Bangladesh and Pakistan are also showing some zeal towards private sector participation in public infrastructure delivery. The PPP market in India has not been very vibrant since 1990s; it was until 2002 that some signs of progress were recorded. Currently, the Indian government has put in place measures and policies including the Viability Gap Fund (VGF) and Project Development Fund (PDF) to support PPP projects initiated in the country (ADB 2011b). Although a central PPP law has not been enacted by the Indian government yet, other states have established legislation and policies to govern the procurement of projects through PPP schemes (ADB 2011b). Considering the strong commitment and positive attitude by the Indian government, it is hoped that the number of projects could increase over years to foster infrastructure growth and development in the country.

In Sub-Saharan Africa (SSA) and Middle East and North Africa (MENA) regions, private investments in PPPs have not been impressive compared with neighbouring regions. In essence, over the past two decades SSA recorded only 509 PPP projects worth US$ 157, 148 million (World Bank XE "World Bank" 2015e). Similarly, MENA recorded 164 PPP projects worth US$ 105, 499 million (World Bank XE "World Bank" 2015e). In SSA, South Africa has contributed most of the region's total account of PPP investments, with 18 out of 27 PPP projects reaching financial close in 2012 (World Bank XE "World Bank" 2013f). South Africa revitalized its PPP market in the early 2000s with a national PPP guideline. The streamlined of approval process coupled with strong local financial market has indeed contributed to some successful PPP projects in South Africa (Osei-Kyei and Chan 2015). Therefore, it is expected that other South African peers within the SSA region would emulate the South African's experience in PPPs to enhance the growth of their local PPP markets.

2.5 Experience of International Case Studies

This section discusses the experience of high-profile case studies from the developed and developing economies. Specifically, cases are selected from Hong Kong and Africa. The selection of a case study was based on two pre-defined criteria: 1) it has to be the first successfully implemented or failed construction PPP project in an infrastructure sector in a country; and 2) it should have readily available information regarding the project's experience and structure (Ameyaw 2015b; Osei-Kyei and Chan 2015, 2016b). All the cases considered in this study satisfied all the above selection criteria, hence were adequately suitable for analysis.

Case studies were analysed in two stages. First, a large range of data and information with respect to projects' performances, structures and features were drawn from governments' reports and documentations, project parties and stakeholders' websites and reports, media commentaries, business magazines, World Bank reports, interviews and articles in newspapers and projects' related literature. Subsequently, the contents of the information and data drawn were thoroughly analysed to derive themes which accurately describes the experiences of the case studies (Osei-Kyei and Chan 2016b; Cheung and Chan 2009b; Kumaraswamy and Zhang 2001b). Follow-up email correspondences with some project stakeholders were done to gain further clarification and insight on projects performances. Considering the above criteria, five high-profiled cases were selected from Nigeria, South Africa, Hong Kong and Ghana. These case studies include, the Lekki Toll Road Project, Inkosi Albert Luthuli Central Hospital, The Hong Kong Cross Harbour Tunnel, Hong Kong Disneyland Theme Park and Ghana National Housing Project.

2.5.1 Lekki Toll Road Concession Project (Nigeria) (Case 1)

2.5.1.1 Background of Project

Over the past decade, transport infrastructure has contributed substantially to the growth of Nigeria's real Gross Domestic Product (Olufowose 2008b). Nevertheless, most transport facilities mainly roads are in deplorable conditions which require urgent upgrade to foster economic activities in Africa's most populous country (Oluwasanmi and Ogidi 2014b). Like other governments, PPP has been considered by the Nigerian government as an effective procurement approach in providing efficient and reliable transport infra- structure system in the country (Oyedele 2014b). The Lekki Toll Road Concession project is the pioneering PPP road project in Nigeria, which was signed between the Lagos State government and a private consortium: the Lekki Concession Company Limited (LCC) (Oxford Business XE "Business" Group 2012b). The private toll road project signed in 2006 primarily sought to improve the deplorable condition and reduce the heavy traffic congestion on the existing road from Lekki to Epe in the Lagos State of Nigeria (World Economic XE "Economics" Forum 2010b).

In the contractual arrangement, the project was to be executed under a Design Build Operate Transfer (DBOT) and Rehabilitate Operate Transfer (ROT) PPP models (ADBG 2008b). The ROT model was applied at the first phase of the project, where the Project Company upgraded and rehabilitated the existing 49.5 km Epe Long Expressway. The DBOT model is yet to be used for the second phase of the project, which would require the consortium to construct and operate the 20 km coastal road on the Lekki Peninsula with an option to improve an existing bypass (LCC 2015b). The project was estimated at an approximately cost of US$ 347 million with a 30-year concession period of which the major source of income is from toll charges (Oforiokuma 2011b; World Economic XE "Economics" Forum 2010b). From a mix of equity and debt capital, funds were raised for the project with the government providing some financial support (ICRC 2012b).

The Lekki Concession project is one of the major transport PPP landmarks in the SSA region, which presents unique experiences that are beneficial to other governments within the region.

2.5.1.2 Project Experience and performance (Success and Failure Factors)

The progress of the Lekki toll road project could heavily be attributed to the emphatic support and commitment demonstrated by the Lagos State government (World Economic XE "Economics" Forum 2010b). In actual fact, the government showed a high level of commitment towards the development of the project by offering a 20-year mezzanine loan of US$ 34 million to the Project Company (ICRC 2012b; World Economic XE "Economics" Forum 2010b). This enabled the concession company to raise substantial funds towards the execution of the project. More importantly, the change of Lagos State government administration did not affect the progress of the project. Although the project was initiated by the previous Lagos State government in 2004, the new government in 2007 ensured that the project was actually implemented. In addition to the financial and political backing, land titles and rights were quickly issued without any delay to the Project Company for the start of the project. The unavailability of land for project development could be a major setback to the progress of PPP deals especially in developing countries (Kumaraswamy and Zhang 2001b); however, the Lekki Concession project did not encounter such difficulty.

The participation of more local content in the Lekki concession project also contributes significantly to its achievements as one of the best toll road projects in the Sub-Saharan region (Lagos State PPP XE "Public-Private Partnership (PPP)" Office 2010b). Essentially, a high pro- portion of the project funds was raised from locally based financial institutions and lenders (ICRC 2012b). Additionally, the concession company had majority of its stakeholders, including the construction firm from within the country which is quite unusual with most PPP projects developed in SSA (Trinity International XE "International" LLP 2009b). The effect of such involvement was that it smoothened and facilitated the construction process as the local construction firm was conversant with the construction management practice within the locality. Moreover many jobs were created with over 635 short-term and 1146 long-term jobs within the locality (World Economic XE "Economics" Forum 2010b).

Although the Lekki road project has been considered as one of Africa's best transport private investment (Lagos State PPP XE "Public-Private Partnership (PPP)" Office 2010b), lots of drawbacks are realised since its operation in 2011. The controversies characterised with the project actually call for critical examination to the development of transport PPPs within the SSA region. The management of the PPP arrangement has over the years been subjected to both political and public scrutiny (Amadi et al. 2014b). The major concern raised on the transport facility is the user fee charges. The toll fees have been a subject of matter to users for the past five years of its operation. Social activists groups and political commentators have on several occasions opposed the toll fees charged by the concessionaire for three different toll plazas on the entire road (Gboyega 2013b). Although such toll collection mechanism adopted was a quick way through which the private consortium could recoup its investment cost, proper consultations were not done with users at the initial stages to avoid such public grievances and oppositions. The stiff opposition from the public forced the State government to adopt a shadow tolling mechanism at the early stage of operation, which was not part of the earlier agreement by parties (O'Karo 2011b). However, the government upon realising the huge cost involved in adopting the shadow toll mechanism quickly reverted to the direct user charges (Nwangwu 2013b). Given the persistent com plaints of high toll charges by commuters and road users, the State government has now considered to acquire the concession rights of the private consortium so as to control the toll fees (Gboyega 2013b). However, such initiative by the government will require a huge amount from the public purse.

In fact, it is unfortunate that as PPPs are seen to relieve governments' budgetary constraints; the Lekki toll road project has rather incurred more financial stress to the public purse due to the grievances expressed by users over tolls. Notwithstanding, the high toll charges and the parties' decision to set three toll plazas on the entire stretch to generate sufficient revenue could be attributed to the high cost of construction and depreciation of the Nigerian Naira (NGN) currency over time. Some social commentators have even asserted that the cost involved in constructing the Lekki road project is outrageous when compared with similar road projects constructed by other State governments in the country (Gboyega 2013b). Nonetheless the Lekki PPP project was not subjected to any competitive tendering process (Aurecon 2014b). This may have limited the public authority's capacity to compare proposals in order to select the most cost-effective bid so as to avoid a huge investment cost, which contributed significantly to the rise in toll fees (Gboyega 2013b).

2.5.2 Inkosi Albert Luthuli Central Hospital (IALCH) (South Africa) (Case 2)

2.5.2.1 Background of Project

In South Africa, the enactment of the Treasury regulation 16 for PPPs in 2000 led to the establishment of the National Treasury's PPP unit (Yong 2010b). This was an attempt by the government to revitalize PPP development after attaining independence in 1994. The IALCH is the first hospital PPP project to be initiated under the established PPP unit in South Africa (Yescombe 2011b). Prior to the completion of the facility, the provincial health department authority had it very difficult to replace old equipment used for services in the hospital due to the limited funds. Moreover, the health authority had wanted to focus on the core functions of the hospital, which is to deliver clinical services and outlet the non – clinical services to the private sector (USAID 2008b). South African's first hospital PPP was signed in 2001 between the Kwa Zulu Natal Department of Health (KZN DoH) and the Impilo private consortium (Farquharson et al. 2011b). The contract required the private consortium to upgrade and manage an 846- bed hospital with the state-of-the-art technology to serve a population of approximately nine million at a total cost of R$ 746 million.

The private partner was to operate the facility and focus more on the non-clinical services for a 15-year concession period of which payment will be received solely based on the availability of the facility (Farquharson et al. 2011b). The hospital project was funded through a mix of equity and long-term debts with the major financier been Rand Merchant Bank Ltd.; a local financial institution. The hospital was opened in 2002 to its first patients.

2.5.2.2 Project Experience (Success and Failure Factors)

In 2002, the hospital PPP project was awarded the best health care PPP deal in Africa (USAID 2008b). So far, project's parties and stakeholders are highly impressed with the performance of the project (Annual report 2008b). The successful progress of the project could be attributed to the commitment and passion shown by the provincial health department (KZN DoH) and the National Treasury's PPP unit. Very often, different governments have different ways of showing commitment towards a PPP project. In the case of IALCH, the KZN DoH demonstrated commitment by providing financial support towards the development of the project. In essence, the KZN DoH contributed an amount of R$360 million which enabled the private consortium to raise substantial funds for the project. Further, the newly instated Treasury's PPP unit also ensured that all approval process had been streamlined to allow the easy flow of the procurement process (Wits Business XE "Business" School 2007c). The streamline of approval process in IALCH was not surprising because it was the first PPP project to be initiated after the establishment of the national PPP unit. Thus, it was obvious that Treasury's approvals could go through with ease (Wits Business XE "Business" School 2007d). As required by the Treasury legislation 16, the hospital project was also subjected to a competitive and transparent tendering process which aided the public health authority's to select a cost effective bid and competent private partner. Since the hospital began operation, stakeholders have been impressed with the private consortium's performance on the service delivery and management (Farquharson et al. 2011b). Importantly, the competition has helped the public client to select a value for money bid.

The competent Transaction Advisor selected by the public authority helped the public authorities to conduct a more detailed feasibility study. Essentially, at the time the hospital project was initiated, South Africa did not have much experience in social infrastructure PPPs; in this regard it was important for an international Transaction Advisor with much experience in hospital PPPs to be engaged. This enabled the KZN DoH to conduct detailed feasibility and to negotiate on the most cost-effective bid for the project (Wits Business XE "Business" School 2007c).

Initially, affordability of hospital services was a great concern to commuters and stakeholders. However, the financial support offered by KZN DoH has avoided high user fees which could have occurred due to exchange rate fluctuations. A much affordable user fee is charged by the hospital authorities so that patients could patronize. In addition, the constant communication has helped sensitized local residents and commuters. Initially, there were agitations on the engagement of the private sector in public health care delivery. However, because of the continuous communication, the general public now has good relationship with the project parties.

2.5.3 *Cross Harbour Tunnel (CHT) (Hong Kong) (Case 3)*

2.5.3.1 Background of Project

The Cross Harbour Tunnel (CHT) is Hong Kong's first BOT project initiated in 1969. This project was signed between the Hong Kong government and the Cross Harbour Tunnel Company Limited (Legislative Council 1999b). The private partner was to build and operate the underwater tunnel project over a 30-year concession period after which the facility will be transferred to the government (Tam 1999b).

The project completed in 1972, cost approximately HK$ 350 million (USD 56 million) (Kumaraswamy and Zhang 2001b). The four lane tunnel of 1852 meters was constructed across the Victoria Harbour; spanning between Hong Kong Island and Kowloon Peninsula. Essentially, before the construction of the tunnel, vehicles and passengers had to travel between the two districts with ferry, therefore it was in no doubt that the completion of the tunnel would enhance economic activities between the two districts (Cheung 2009b).

2.5.3.2 Project Experience

Ideally all available information and documentaries on the CHT indicate that the project was successful (Lee 2005b). The success of Hong Kong's first BOT tunnel could be attributed to the strong commitment and support shown by the government towards the development of the tunnel project. In essence, the government demonstrated commitment towards the project by retaining a percentage of the Project Company's share (Tam 1999b). The government actually owned 20% share of the Project Company, which in a way restricted the government from creating competing projects for the CHT. Very often, when an alternative or a competing project is available, it increases the other project's demand risk. Moreover, revenue collection slows down due to competition and this affects revenue targets. This even becomes worse when the alternative project is provided freely to users. In the case of CHT project because the government had a percentage of commitment in the project, no alternative was made available (Tam 1999b) and the project was delivered and operated effectively.

After few years of opening, the traffic flow of the tunnel project increased from 20,000 to 124,000 vehicles per day, this truly demonstrated the high demand for the CHT project (Tam 1999b). It was therefore unsurprising that the Project Company recouped its investment cost after only three and half years of operation (Cheung et al. 2009b). Some commentators have attributed the increased traffic flow or demand to the location of the project and the rapid growth of Hong Kong's economy in the late 1960s and 70s (Cheung 2009b; Mak and Mo 2005b). Notwithstanding, PPP projects must be a necessity and they must have a long term demand to ensure a continuous income stream to the Project Company as realised with the CHT project.

The sophisticated technology and innovation employed in the CHT by the Project Company contributed greatly to its success (Tam 1999b). During the unexpected increase in traffic flow, an electric toll collection system was used to reduce congestion at toll booths. This however allowed a free flow of traffic, which help minimized public agitations on traffic congestions. Another innovative method adopted is the immerse tube technique used for the construction of the underwater tunnel (Mak and Mo 2005b). This technology contributed to the fast delivery of the project even before the anticipated construction period.

2.5.4 *Hong Kong Disneyland Theme Park (Hong Kong) (Case 4)*

2.5.4.1 Background of Project

Another promising PPP project in Hong Kong is the Hong Kong Disneyland Theme Park. This project is also one of the initiated and well implemented PPP projects which have made significant contribution to Hong Kong's economy and tourism industry (Shen et al. 2006b). The Hong Kong Disneyland is a Joint Venture agreement signed between the Hong Kong government and Walt Disney Company in 1999, for the creation of a theme park and resort complex (Matusitz 2011b). Parties to the joint venture provided equity financing towards the development of the project. The Hong Kong government contributed HK$ 3.25 billion owing 57% shares while Disney Company invested HK$2.45 billion with 43% shares (Esty 2001b). Moreover, in the partnership agreement, the Hong Kong government was responsible for land reclamation and other infrastructure facilities while the private partner was also responsible for the construction and operation of the resort (Leung and Hui 2002b). The Hong Kong Disneyland project was constructed on the Lantau Island and was finally opened for services in 2005 (Shen et al. 2006b).

2.5.4.2 Project Experience

Considering the successful progress and significant contribution of the Disney PPP project to Hong Kong's economy and international reputation, it is regarded as one of Hong Kong's landmark PPP projects particularly in the tourism sector (HKDRMR 2014b). One contributing factor towards the successful progress of the Theme Park is the commitment demonstrated by the government (Matusitz 2011b). In spite of the many public disparagements concerning the huge investment cost borne by the government in the joint venture agreement; the Hong Kong government still believed that the economic benefits expected from the project are huge. Hence there was the need to fully commit and support this first Disney project in Hong Kong (Matusitz 2011b). The emphatic support and commitment from the government has indeed made the Disney Theme Park a reality and a key contributor to Hong Kong's Gross Domestic Product (HKDRMR 2014b).

The proper management techniques and innovations employed during the operational stage have also contributed to the success and achievements of the Theme Park project. Since the Hong Kong Disneyland started operating in 2005, it has been competing with other existing Theme Parks in Hong Kong's tourism industry particularly the Ocean Park located at the Southern District of Hong Kong (Huang and Hsu 2005b). As usual, the existence of competing projects did not allow the Disneyland project to meet its revenue target at the early stages of operation. However, in order to adequately compete with other tourist attractions and sites, the management of the facility employed proper measures and innovations such as the incorporation of more local customs and culture in their operations to attract more local tourist than previous (Matusitz 2011b). After employing such measures, the number of local tourists increased, meeting the anticipated tourist target (HKDRMR 2014b). Another contributing factor towards the successful delivery of the Theme Park project is the proper allocation of risks among parties (Shen et al. 2006b). Before Hong Kong government entered into the agreement, concerns were raised with respect to the risky nature of Theme Park projects drawing on the performance of the Disneyland Paris in France. However, in the Hong Kong Disneyland project, risks including site acquisition, land reclamation and financial risks were allocated to the party with better mitigation techniques. The proper allocation and complete transfer of risks has increased the project's management efficiency since its inception (Shen et al. 2006b; Leung and Hui 2002b). Essentially, the Hong Kong Disneyland Theme Park symbolizes a very well-planned tourist attraction PPP project which offers good prospects to Hong Kong's tourism industry. In this regard the lessons presented will be beneficial for future implementation of similar PPP projects.

2.5.5 The Ghana National Housing Project (GNHP) (Ghana) (Case 5)

2.5.5.1 Background of Project

The GNHP was part of the affordable housing policy initiative by GoG in 2005. It was the first ever major PPP housing project to be initiated in the country's housing sector since mid-1960s (Mingle 2010b). It was a partnership deal signed between the GoG through its MWRWH and a Ghana- Korea private developer (STX Engineering and Construction Ghana Ltd). The housing project agreement was signed in 2009. The project was to provide 200,000 housing units in the ten regional capitals of Ghana (Oxford Business XE "Business" Group 2011d) and the estimated cost of the project was approximately US$ 10 billion, which was expected to be constructed in 4 phases starting from 2010.

The first phase involved the construction of 30,000 housing units, followed by 60,000 for the second phase. The remaining 110,000 were to be constructed over the remaining two phases. Out of the 200,000 housing units, 30,000 (first phase) was expected to be completed within 5 years spanning from 2010 to 2015 at approximately US$ 1.3 billion excluding taxes and duties; whereas the remaining 170,000 units were to be completed over the years. In fact, the government intended to allocate the first phase of the housing units to the security services across the ten regions of Ghana (Memorandum of Understanding 2010b).

The GoG and the STX Engineering and Construction Ghana Ltd. entered into an off-take arrangement for the project (Memorandum of Understanding 2010b). As agreed between the parties, the GoG was expected to purchase 90,000 (45%) of the 200,000 housing units constructed by the private developer (Off – Taker agreement 2009). The remaining 110,000 (55%) housing units would be off take by a local mortgage financial institution (i.e HFC bank). However, the repayment of the off-take for the 45% to the private developer was expected to take 15 years (African Elections Project XE "Projects" 2010b). The contractual agreement clearly spelt out the roles of the parties in the partnership. The private developer was obliged to;

i. Plan, design and construct the housing project according to international standards on the available sites
ii. Obtain funds for the construction of the housing project with the financial guarantee by the government.
iii. To utilize the local resources of Ghana with over 30% for executing the housing project and as well provide some training to local labour force employed in the project delivery.

Likewise the public partner (i.e GoG) was to:

i. Make lands available for the construction of the project at no cost to the investor.
ii. Make available all necessary utilities such as water, electricity, roads, and any other relevant facility for the delivery of the project.
iii. Provide tax exemptions on incomes of sub-contractors and expatriate personnel, import duties on equipment, materials as well as machinery for the purpose of the housing delivery.
iv. Facilitate the processing of any approvals or granting of permits in relation to the project delivery (Off – Taker agreement 2009).

2.5.5.2 The Project Finance

As mentioned earlier, the estimated cost of the 200,000 housing unit was US$10 billion with the first 30,000 housing unit estimated to be approximately US$1.3 billion excluding taxes and duties. In order to help the private developer, raise adequate funds for the project, the government provided a sovereign guarantee. It was expected that with the sovereign guarantee the private developer could raise funds for the housing project from the global financial market. The sovereign guarantee had a 20-year maturity period.

It should be noted that the sovereign guarantee was given as part of the conditions on the off-take agreement between the parties. In this regard the sovereign guarantee covered only the 90,000 (45%) housing units to be off take by the GoG. Further, it was still expected that the Korean Government through its Global Infra Fund (a sovereign fund established by the South Korean government to finance international infrastructure projects by South Korean firms) would be used as well as the collaborative efforts from other local and international financial institutions to support the private investor in financing the remaining 55% (110,000) housing units (Oxford Business XE "Business" Group 2011c; Africa Election Project 2010b).

2.5.5.3 Project Experience

Five underlying factors are identified to have contributed to the failure of the GNHP. These include weak private consortium structure, strong political opposition, absence of competitive procurement process, lack of experience in handling housing PPP deals and Unavailability of funds for project delivery

- ***Weak private consortium structure***

The long-term and complex structure of PPP projects requires a very strong and potential project company for their successful delivery (Cheung et al. 2012). On the contrary, a weak multidisciplinary team of stakeholders and private partners in the Special Purpose Vehicle (SPV) would not be capable to successfully undertake such a complex and long term nature project (Chan et al. 2010d).

The contention within the boardroom of the project company contributed to the failure of the housing PPP (Owusu 2011b). The private consortium was a joint venture between a Ghanaian private company (G.K Airports Company Limited) and a South Korean construction firm (STX Construction Group) (Citifmonline 2011b). These companies formed a consortium (STX Engineering and Construction Ghana Limited) for the purpose of undertaking the GNHP. However, after the signing of GNHP contract with the government and sod cutting to begin construction, a board room wrangling emerged between the local and foreign companies. The two parties began arguing over the full ownership right of the project company.

In fact, this in-house fighting ended up in an intense legal battle between the private parties (Dzawu 2012b). Major concerns raised by the parent partner (South Korean STX Construction Group) of the SPV was that the Ghanaian representative of the Project Company (STX Engineering and Construction Limited) had sidelined the parent partner during negotiations of the partnership agreement with the government (Benson 2011b). The South Korean partner claimed that their Ghanaian partner had not correlated with them properly on some of the details of the partnership agreement that was signed on behalf of the Project Company. They further asserted that most of the details of the partnership agreement were known through media reports. In addition, the parent partner claimed that their Ghanaian partner undertook some actions regarding the proposed project without their consents and notice (Dzawu 2012b). These claims by the South Korean partners elicited them to seek legal action to determine who owns the SPV and therefore the appropriate partner to execute the project. On the other hand, the Ghanaian partner also sorted a legal action for the court to declare them as the lawful partner to execute the GNHP (Benson 2011b).

This boardroom wrangling between the private parties delayed the commencement of the project close to six months after agreements have been signed. The government realizing the conflicts decided to back out from the partnership agreement and began the due process to retrieve all national assets provided for the GNHP (Dzawu 2012b). The GNHP experience actually reinforces the assertion made that a strong multi-disciplinary consortium is very key in PPPs especially in developing countries where often local private developers partner foreign companies to undertake a housing PPP project.

- ***Strong political opposition***

In Ghana's political setting, opposition political parties play crucial role in shaping incumbent government's policies and initiated project deals, therefore most projects initiated by existing government are highly assessed and of great concern to the opposition parties (Mustapha and Whitfield 2009b). The GNHP received strong resistance from the opposition political parties. Although the opposition parties claimed they were not against reducing the increasing housing deficit of 1.7 million, their major concern was on the nature and structure of the housing project contract, and its ability to provide value for money for the public (Gyasiwaa 2011b).

The intense opposition contributed to delays in the agreement of certain conditions raised by political opponents in the contract. One key issue raised by the major opposition party was the numerous incentives given to the private developer. Other issues include tax exemptions on incomes of expatriate personnel and subcontractors, exemptions of import duties on machinery, equipment and materials. In addition, the opposition parties lamented that a due diligence had not been conducted in awarding the contract. Moreover, they indicated that the government had agreed on a high interest rate and payment fees, which certainly are in favour of the private developer (GNA 2010b).

Essentially, the stiff opposition from the political front increased the general public's perception on the GNHP as fraudulent and not beneficial. Notwithstanding, despite the agitation by the opposition party, some section of Ghanaians still believed that the GNHP is beneficial not only in addressing the housing deficit, but it also strengthens the Ghanaian housing market due to its first kind (Mingle 2010b).

- ***Absence of competitive procurement process***

The GNHP was not subjected to any competitive bidding process (Amesimeku 2013b) but rather, the project was introduced to the government by one of the directors of the Project Company (Benson 2011b). This therefore raised lots of arguments and concerns about the estimated cost of the project and end user charges (Daily Guide 2010b). Most professional bodies and policy experts emphasized that the project could be delivered at a much lower cost than the estimated cost by the selected private developer (GNA 2010b). Moreover, with the numerous incentives and strong commitment from the government, many believed that given such support and opportunity, a competitive procurement process would have allowed the government to select a capable private investor with much lower estimated cost compared to the one agreed on by the government (Oxford Business XE "Business" Group 2011d). Owing to this, it is realized that the absence of competition contributed to such high estimated cost of the GNHP and attracted public scrutiny and criticism. Indeed, GNHP affirms the assertion that the lack of competition in PPP procurement process results in high project cost and the absence of value for money assurance (Chan et al. 2010c).

- ***Lack of experience in handling housing PPP deals***

PPP as a procurement approach is a growing concept globally and it requires experience and adequate understanding on how this concept operates to ensure its effective implementation (Li et al. 2005). PPP cannot be successfully implemented overnight; it requires training to enhance public and private sectors' management skills (Morledge and Owen 1998b). Since its evolution in Ghana for construction projects in 2004, this concept is still developing and has not yet been fully explored and understood especially in the Ghanaian housing sector (UN-Habitat 2009b). It will therefore require adequate capacity building to fully understand how this new procurement model operates within the local context.

The GNHP contract was poorly managed by the public and private sectors. Essentially, most contractual conditions were not properly and clearly agreed on by parties. Additionally, issues pointed out by external stakeholders (public/ users, opposition parties and professional bodies) were not properly addressed to allow the easy flow of the project. In some situations, some contract conditions and agreements were withdrawn from the Parliament of Ghana to be reconsidered by the project parties (GNA 2010b). This resulted into suspicions of corrupt practices by the general public. This therefore contributed to the negative public perceptions on the GNHP deal.

- ***Unavailability of funds for project delivery***

Extant literature has emphasised that the private sectors' ability to raise massive funds for public projects is one of the major reasons why most governments engage in PPPs (Walker and Smith 1995b; Osei-Kyei et al. 2014b). Indeed, this assertion did not really manifest in the GNHP. Although the government provided a sovereign guarantee to enable the Project Company raise funds, it was still challenging for the Project Company to raise substantial funds for the delivery of the project (Owusu 2011b).

Given the huge cost of project and unclear terms and conditions of the contract, it was difficult for international financial institutions to provide credit. Moreover, the huge controversies surrounding the project made some international financial institutions reluctant to invest in the facility (Statesman 2011b). These controversies considerably increased the financial risk of the facility particularly when these controversies emanated from political fronts and users/ public.

2.6 Chapter Summary

This chapter has explored the evolution of the PPP concept. Further the development of PPP from developed and developed economies have been discussed. In the developed economies, countries including the U.K., Australia, Canada and the U.S have developed their PPP markets, whereas in developing economies countries such as China, India, Brazil and South Africa have made significant progress in the implementation of the PPP concept. The chapter has also explored the experience of high-profiled cases from developing and developed economies. The cases discussed include the Lekki Toll Road Project, Inkosi Albert Luthuli Central Hospital, The Hong Kong Cross Harbour Tunnel, Hong Kong Disneyland Theme Park and Ghana National Housing Project.

References

African Development Bank Group (ADBG). (2008b). *Nigeria: AfDB Approves US$ 85 Million for Lekki Toll Road Project*. Available from: http://www.afdb.org/en/news-and-events/article/nigeria-afdb-approves-us-85-million-for-lekki-toll-road. Accessed on 31 Dec 2014.

African Elections Project. (2010b). *Minister outdoors plan for STX Project*. Available from: http://www.africanelections.org//ghana/postelections/news/page.php?news=451. Accessed on 7 Nov 2014.

Akintoye, A., Beck, M., & Hardcastle, C. (2008b). *Public-private partnerships: Managing risks and opportunities*. Blackwell Science.

Amadi, C., Carrillo, P., & Tuuli, M. (2014b). Stakeholder management in public private partnership projects in Nigeria: Towards a research agenda. In A. B. Raiden & E. Aboagye-Nimo (Eds.), *Proceedings of the 30th annual ARCOM conference, 1–3 September 2014, Portsmouth, UK*.

Amesimeku E. (2013b). Ghanaians express cautious optimism on new affordable housing project. *Property Investor Africa*, 23 December. Available from: https://propertyinvestorafrica.com/ghanaians-express-cautious-optimism-on-new-affordable-housing-project. Accessed on 1 Nov 2014.

Ameyaw E.E (2015b). *Risk allocation model for Public-Private Partnership water supply projects in Ghana* (Doctoral dissertation, The Hong Kong Polytechnic University).

Annual Report. (2008b). *Inkosi Albert Luthuli Central Hospital*. Available from: http://www.ialch.co.za/index.php/resources/annual-reports?download=3:annual-report-2007-2008. Accessed Feb 2015.

Appleby, G., Aroney, N., & John, T. (2012). *The future of Australian federalism: comparative and interdisciplinary perspectives*. Cambridge University Press.

Asian Development Bank (ADB). (2011b). *Public-private partnership infrastructure projects: Case studies from the Republic of Korea*. Available from: http://ppp.worldbank.org/public-private-partnership/library/public-private-partnership-infrastructure-projects-case-studies-republic-korea-0. Accessed on May 2015.

Aurecon. (2014b). *Lekki-Epe Expressway Toll Road, Nigeria. Aurecon Group*. Available from: http://www.aurecongroup.com/en/projects/transport/lekki-epe-expressway-toll-road. Accessed on 4 Jan 2015.

Beauregard, R. A. (1998b). Public-private partnerships as historical chameleons: the case of the United States. In J. Pirre (Ed.), *Partnerships in urban governance*. London: MacMillan.

Benson I. (2011b). STX Deal to start at last. *The Chronicle*, November 16. Available from: http://thechronicle.com.gh/stx-deal-to-start-at-last/. Accessed on 23 Oct 2014.

Bovis, C. (2013b). *Public-private Partnerships in the European Union*. London: Routledge.

Canadian Council for Public Private Partnerships (CCPPP). (2013b). *10 – year economic impact assessment of PPP in Canada 2003–2012. PPP Canada*. Available from: http://www.p3canada.ca/en/about-p3s/p3-resource-library/10-year-economic-impact-assessment-of-public-private-partnerships-in-canada. Accessed May 2015.

Chan, A. P. C., & Cheung, E. (2014b). *Public private partnerships in international construction: Learning from case studies*. New York: Routledge.

Chan, A. P., Lam, P. T., Chan, D. W., Cheung, E., & Ke, Y. (2010c). Potential obstacles to successful implementation of public-private partnerships in Beijing and the Hong Kong special administrative region. *Journal of Management in Engineering*. https://doi.org/10.1061/(ASCE)0742-597X(2010)26:1(30).

Chan, A. P., Lam, P. T., Chan, D. W., Cheung, E., & Ke, Y. (2010d). Critical success factors for PPPs in infrastructure developments: Chinese perspective. *Journal of Construction Engineering and Management, 136*(5), 484–494.

Cheung, E. (2009b). *Developing a best practice framework for implementing public private partnerships (PPP) in Hong Kong* (Doctoral dissertation, Queensland University of Technology).

Cheung, E., & Chan, A. P. C. (2009b). Is BOT the best financing model to procure infrastructure projects?: A case study of the Hong Kong-Zhuhai-Macau Bridge. *Journal of Property Investment & Finance, 27*(3), 290–302.

Cheung, E., Chan, A. P., & Kajewski, S. (2009b). Reasons for implementing public private partnership projects: Perspectives from Hong Kong, Australian and British practitioners. *Journal of Property Investment & Finance, 27*(1), 81–95.

Cheung, E., Chan, A. P. C., & Kajewski, S. (2012). Factors contributing to successful public private partnership projects, Comparing Hong Kong with Australia and the United Kingdom. *Journal of Facilities Management, 10*(1), 45–58.

Citifmonline. (2011b). *BK Asamoah: I'm still STX Ghana Boss*. Available from: http://www.ghanaweb.com/GhanaHomePage/NewsArchive/artikel.php?ID=216675. Accessed on 2 Oct 2014.

Corrigan, M. B., Hambene, J., Hudnut, W., III, Levitt, R. L., Stainback, J., Ward, R., & Witenstein, N. (2005b). *Ten principles for successful public/private partnerships*. Washinton, DC: Urban Land Institute.

Daily Guide. (2010b). *GREDA chickens out of opposition to STX housing deal*. Available from: http://www.modernghana.com/news/283489/1/greda-chickens-out-of-opposition-to-stx-housing-de.html. Accessed on 16 Nov 2014.

Dzawu, M. M. (2012b). Ghana's capital dries up as water system lags oil-led boom. *Bloomsberg Business*. 4th April. Available from: http://www.bloomberg.com/news/articles/2013-04-04/ghana-s-capital-dries-up-as-water-system-lags-oil. Accessed on 11 Jan 2015.

Eggers and Startup. (2006b). *Closing the infrastructure gap: The role of PPPs*. Deloitte: Deloitte Research Study.

English, L. M. (2006b). Public private partnerships in Australia: An overview of their nature, purpose, incidence and oversight. *University of New South Wales Law Journal, 29*, 250.

Esty, B. C. (2001b). Structuring loan syndicates: A case study of the Hong Kong Disneyland project loan. *Journal of Applied Corporate Finance, 14*(3), 80–95.

Farquharson, E., de Mästle, C. T., & Yescombe, E. R. (2011b). *How to engage with the private sector in public-private partnerships in emerging markets*. Washington, DC: World Bank Publications.

Gboyega, A. (2013b). Lekki-Epe Road: Lagos to raise N87.5bn to acquire concession rights. *This Day Live,* 29 August. Available from: http://www.thisdaylive.com/articles/lekki-epe-road-lagos-to-raise-n87-5bn-to-acquire-concession-rights/157543/. Accessed on 2 Jan 2015.

Ghana News Agency. (2010b). *Issues criticized in STX housing deal not addressed – Minority.* Available from: http://www.ghananewsagency.org/print/18673. Accessed on 5 Nov 2014.

Gifford, J. (2013b). *Public private partnerships: Prospect for the future international transport forum, Leipzig, Germany.* Available from: http://www.ppptransport.eu/docs/ITF2013/3_Gifford.pdf. Accessed May 2015.

Gyasiwaa, A. (2011b). *Akyena Brentuo: STX deal will happen.* Myjoyonline, August 22. Available from: http://politics.myjoyonline.com/pages/news/201108/71540.php. Accessed on 4 Nov 2014.

HM Treasury and Infrastructure U.K. (2013b). *PFI projects data: March 2013.* Available from: https://www.gov.uk/government/statistics/pfi-projects-data-march-2013. Aug 2014.

Ho, P. H. (2006b). Development of public private partnerships (PPPs) in China. *International Roundup, 19*(1).

Hong Kong Disneyland Resort Media Relations (HKDRMR). (2014b). *Walt Disney parks and resorts.* Available from http://hkcorporate.hongkongdisneyland.com/hkdlcorp/en_US/home/home?name=HomePage. Accessed Feb 2015.

Huang, S., & Hsu, C. H. C. (2005b). Mainland Chinese residents' perceptions and motivations of visiting Hong Kong: Evidence from focus group interviews. *Asia Pacific Journal of Tourism Research, 10*(2), 191–205.

Infrastructure Concession Regulatory Commission (ICRC). (2012b). *PPP project case studies.* Available from: http://ppptoolkit.icrc.gov.ng/ppp-project-case-studies/. Accessed on 18 Jan 2015.

Jooste, S. F. (2010b). *Exploring the networks of organizations that enable and govern infrastructure public-private partnerships: An organizational field-level perspective.* Stanford: Stanford University.

Kumaraswamy, M. M., & Zhang, X. Q. (2001b). Governmental role in BOT-led infrastructure development. *International Journal of Project Management, 19*(4), 195–205.

Lagos State PPP Office. (2010b). *Lagos Infrastructure Project (LIP) Concession.* Available from: http://www.lagosstateppp.gov.ng/projects/project_portfolio/project_in_construction.asp. Accessed on 20 Jan 2015.

Lee, V. (2005b). *Public – Private Partnerships.* Hong Kong: *Research and Library Services Division Legislative Council Secretariat.*

Legislative Council. (1999b). *Legislative Council Brief, Road Tunnels (Government) Ordinance (Cap. 368).* Hong Kong Government. Accessed from: http://www.legco.gov.hk/yr98-99/english/bc/bill_82/general/82_brf.pdf. Accessed on 3 Dec 2014.

Legislative Council Secretariat. (2006b). *The adoption of the build-operate-transfer model and its variants in developing transport infrastructures in selected places research and library services division.* http://www.legco.gov.hk/yr05-06/english/sec/library/0506in13e.pdf. Accessed on 10 Jan 2015.

Lekki Concession Company (LCC). (2015b). *Project scope.* Available from: http://www.lcc.com.ng/tolls.asp?pid=21. Accessed on 18 Jan 2015.

Leung, B. Y., & Hui, E. C. (2002b). Option pricing for real estate development: Hong Kong Disneyland. *Journal of Property Investment & Finance, 20*(6), 473–495.

Levy, S. M. (1996b). *Build, operate, transfer: Paving the way for tomorrow's infrastructure.* New York: Wiley.

Li, B. (2003b). *Risk management of construction public private partnership projects.* Ph.D. thesis, Glasgow Caledonian University, UK.

Li, B., Akintoye, A., Edwards, P. J., & Hardcastle, C. (2005). Critical success factors for PPP/PFI projects in the UK construction industry. *Construction Management and Economics, 23*(5), 459–471.

Mak, C. K., & Mo, S. (2005b). *Some aspects of the PPP approach to transport infrastructure development in Hong Kong*. Highways Department, Government of the Hong Kong Special Administrative Region.
Matusitz, J. (2011b). Disney's successful adaptation in Hong Kong: A glocalization perspective. *Asia Pacific Journal of Management, 28*, 667–681.
Memorandum of Understanding (MOU). (2010b). Available from: http://danquahinstitute.org/docs/STX-MOU.pdf. Accessed on 17 Nov 2014.
Meng, X., Zhao, Q., & Shen, Q. (2011b). Critical success factors for transfer-operate-transfer urban water supply projects in China. *Journal of Management in Engineering*. https://doi.org/10.1061/(ASCE)ME.1943-5479.0000058.
Mingle J. (2010b). *Commentary on STX housing project*. Ghana Broadcasting Corporation. Available from: http://gbcghana.com/kitnes/cache/dynamic/1156360.htm. Accessed on 25 Oct 2014.
Morledge, R., & Owen, K. (1998b). *Critical success factors in PFI projects*. Proceedings of the 14th ARCOM Annual Conference, (AAC' 98), University of Reading, pp. 565–574.
Mustapha, A. R., & Whitfield, L. (2009b). *Turning points in African democracy*. Rochester: Boydell & Brewer.
Nwangwu, G. A. (2013b). *A risk based approach to enhancing public-private partnership (PPP) projects in Nigeria*. PhD thesis, University of Hull, UK.
O'Karo T. J. (2011b). *The Lekki Toll Gate Protest – An eyewitness account*. Bella Naija, 19 December. Available from: http://www.bellanaija.com/2011/12/19/the-lekki-toll-gate-protest-an-eyewitness-account/. Accessed on 10 Dec 2014.
Off-Teaker Agreement. (2009). Available from: danquahinstitute.org/docs/STX-Off-TakerAgreement.pdf. Accessed on 12 Nov 2014.
Oforiokuma O. (2011b). *PPPs and sub nationals – Lessons from around the World: Lekki Toll Road Concession, Lagos, Nigeria*. IFC Regional PPP Conference 14th November 2011. Available from: http://www.ifc.org/wps/wcm/connect/345f92004956d9af933bbf849537832d/9.b2_CaseStudiesTollRoads-LekkiConcession.pdf?MOD=AJPERES. Accessed on 15 Jan 2015.
Olufowose, E. (2008b). *Infrastructure finance in Nigeria: Air and water transport*. 2nd annual banking and finance conference of the Chartered Institute of Bankers of Nigeria (CIBN) 28th–30th October, Abuja, pp. 1–3.
Oluwasanmi, O., & Ogidi, O. (2014b). Public private partnership and Nigerian economic growth: Problems and prospects. *International Journal of Business and Social Science, 5*(11).
Organisation for Economic Co-operation and Development. (2008b). *Public-private partnerships: In pursuit of risk sharing and value for money*. Paris: OECD Publications.
Organisation for Economic Co-operation and Development. (2010b). *Dedicated public-private partnership units: A survey of institutional and governance structures*. Paris: OECD Publishing.
Osborne, S. (2002b). *Public-private partnerships: Theory and practice in international perspective*. New York: Routledge.
Osei-Kyei, R., & Chan, A. P. C. (2015). Developing transport infrastructure in Sub-Saharan Africa through public–private partnerships: Policy practice and implications. *Transport Reviews*, 1–17.
Osei-Kyei, R., & Chan, A. P. C. (2016b). Implementing public–private partnership (PPP) policy for public construction projects in Ghana: critical success factors and policy implications. *International Journal of Construction Management*. https://doi.org/10.1080/15623599.2016.1207865.
Osei-Kyei, R., & Chan, A. P. (2017). Factors attracting private sector investments in public–private partnerships in developing countries. *Journal of Financial Management of Property and Construction*.
Osei-Kyei, R., Dansoh, A., & Ofori–Kuragu, J. K. (2014b). Reasons for adopting Public–Private Partnership (PPP) for construction projects in Ghana. *International Journal of Construction Management, 14*(4), 227–238.
Owen, G., & Merna, A. (1997b). The private finance initiative. *Engineering Construction and Architectural Management, 4*(3), 163–177.

Owusu W. Y. (2011b). STX cheques bounced. *Daily Guide*, 22 August. Available from: http://www.modernghana.com/news/346347/1/stx-cheques-bounced.html. Accessed on 1 Nov 2014.

Oxford Business Group. (2011c). *The report: Ghana 2011*. London: Oxford Business Group Publications.

Oxford Business Group. (2011d). *Ghana: No place like home*. Available from: http://www.oxford-businessgroup.com/news/ghana-no-place-home. Accessed on 3 Oct 2014.

Oxford Business Group. (2012b). *The report: Nigeria 2012*. London: Oxford Business Group Publications.

Oyedele O. A. (2014b). *Public-Private Partnership (PPP) and infrastructure provision in Nigeria*. AARC HES Conference.

Price Waterhouse Coopers. (2012b). *Public-Private Partnerships in Malaysia*. Available from: http://www.pwc.com/my/en/assets/services/ppp-projects-in-malaysia.pdf. Accessed May 2015.

Reinhardt, W. (2011b). The role of private investment in meeting US transportation infrastructure needs. *Public Works Financing, 260*, 1–31.

Shen, L. Y., Platten, A., & Deng, X. P. (2006b). Role of public private partnerships to manage risks in public sector projects in Hong Kong. *International Journal of Project Management, 24*(7), 587–594.

Shimron, B. (1979). Polybius on Rome: a re-examination of the evidence. *SCI5 (1979–80)*, 94–117.

State Government of Victoria. (2015b). *Public-Private Partnership. State of Victoria, Australia*. Available from: http://www.dtf.vic.gov.au/Infrastructure-Delivery/Public-private-partnerships. Accessed Jan 2015.

Statesman. (2011b). *Bagbin: Yes, we have secured STX funds, from Stanchart, no, I mean, Barclays*. Available from: http://www.ghanaweb.com/GhanaHomePage/NewsArchive/artikel.php?ID=213071

Stromback L. (1993b). *Baltzar von Platen, Thomas Telfordoch Göta Kanal: entreprenörskap och tekniköverföring i brytningstid*. Doctoral thesis, Symposion Graduale, Stehag, Sweden.

Tam, C. M. (1999b). Build-operate-transfer model for infrastructure developments in Asia: Reasons for successes and failures. *International Journal of Project Management, 17*(6), 377–382.

Trebilcock, M., & Rosenstock, M. (2013b). Infrastructure PPPs in developing world: Lessons from recent experience mimeo.

Trinity International L.L.P. (2009b). *A review of the Lekki-Epe Expressway Toll Road Concession Project*. Available from: http://www.trinityllp.com/a-review-of-the-lekki-epe-expressway-toll-road-concession-project/. Accessed on 21 Dec 2014.

United Nations Human Settlements Programme (UN – Habitat). (2009b). *The UN-HABITAT slum upgrading facility. Newsletter*, Issue No. 7. Available from: www.mirror.unhabitat.org/pmss/getElectronicVersion.aspx?nr=2639&alt=1. Accessed on 4 Sept 2014.

United States Agency for International Development (USAID). (2008b). *The South African PPP program: The Inkosi Albert Luthuli Hospital Transaction*. Available from: http://www.energytoolbox.org/library/ppp_in_infrastructure/presentations/The_South_Africa_PPP_Program.pdf. Accessed on Feb 2015.

Walker, C., & Smith, A. J. (1995b). *Privatized infrastructure: The BOT approach*. London: Thomas Telford.

Wall, A., & Wall, T. (2013b). *Public-private Partnerships in the USA: Lessons to be Learned for the United Kingdom*. New York: Routledge.

Winch, G. M. (2000b). Innovativeness in British and French construction: the evidence from Trans manche–Link. *Construction Management and Economics, 18*, 807–817.

Wits Business School. (2007c). *The Public Private Partnership between the KwaZulu Natal Department of Health and Impilo Consortium in Inkosi Albert Luthuli Central Hospital*. Available from: http://www.ppp.gov.za/Legal%20Aspects/Case%20Studies/IALCH%20final%20document%20comparison.pdf. Accessed Feb 2015.

Wits Business School. (2007d). *Case studies on the PPPs at Humansdorp District Hospital Universitas and Pelonomi Hospitals and Inkosi Albert Luthuli Central Hospital*. Available from:http://www.ppp.gov.za/Legal%20Aspects/Case%20Studies/Humansdorp%20 Overall%20findings.pdf. Accessed Feb 2015.

World Bank. (2013d). *Infrastructure Policy Unit 2012 Latin America and the Caribbean PPI Data Update, (Note 93)*. Washington, DC: World Bank Publications.

World Bank. (2013e). *Infrastructure Policy Unit 2012 East Asia and the Pacific PPI Data Update. (Note 91)*. Washington, DC: World Bank publications.

World Bank. (2013f). *Infrastructure Policy Unit 2012 Africa PPI Data Update. (Note 90)*. Washington, DC: World Bank publications.

World Bank. (2014). *2014 Global PPI Updates*. Washington, DC: World Bank Publications.

World Bank. (2015d). *Private Participation in Infrastructure (PPI), regional snapshots*. Washington, DC: World Bank. Available from: http://ppi.worldbank.org/explore/ppi_exploreRegion.aspx?regionID=1. Accessed on May 2015.

World Bank. (2015e). *Private Participation in Infrastructure (PPI), Middle East and North Africa. Regional snapshots*. Washington, DC: World Bank. Available from: http://ppi.worldbank.org/explore/ppi_exploreRegion.aspx?regionID=1. Accessed on May 2015.

World Economic Forum. (2010b). *Paving the way: Maximizing the value of private finance in infrastructure*. World Economic Forum, USA. Available from: http://www3.weforum.org/docs/WEF_IV_PavingTheWay_Report_2010.pdf

Yescombe, E. R. (2011b). *Public-private partnerships principles of policy and finance*. Elsevier.

Yong, H. K. (Ed.). (2010b). *Public-private partnerships policy and practice: A reference guide*. Commonwealth Secretariat.

Chapter 3
Reasons for Governments Adoption of Public-Private Partnership

Abstract Following the evolvement of the PPP concept in the early 1990s, governments globally have shown strong enthusiasm towards the adoption of this concept in various infrastructure sectors. It is however not clear as to the motivations of these governments towards the adoption of the PPP concept. Many critics suggest that the main reason is due to the lack of funds, but is that the only reason? Further, different jurisdictions/economies have different conditions and political agendas, therefore it is possible that there could be some differences in the motivations of governments adoption of the PPP concept particularly among developing and developed economies. This chapter seeks to explore the above raised concerns and discuss comprehensively on the differences and similarities in reasons/motivation for PPP adoption by governments in developing and developed economies based on empirical data.

Keywords Reasons for PPP · Motivations · Ghana · Hong Kong · Developed economies · Developing economies

3.1 Introduction

Since the PPP concept evolved in the early 1990s, governments globally have shown strong enthusiasm towards the adoption of this concept in various infrastructure sectors. It is however not clear as to the motivations of these governments towards the adoption of the PPP concept. Many critics suggest that the main reason is due to the lack of funds, but is that the only reason? Further, different jurisdictions/economies have different conditions and political agendas, therefore it is possible that there could be some differences in the motivations of governments adoption of the PPP concept particularly among developing and developed economies. This chapter seeks to explore the above raised concerns and discuss comprehensively on the differences and similarities in reasons/motivation for PPP adoption by governments in developing and developed economies based on empirical data. Osei-Kyei and Chan (2018) conducted a questionnaire survey with experienced PPP practitioners in Ghana and Hong Kong. The selected practitioners were requested to rank the significance of twelve identified set of government motivations for adopting PPP.

R. Osei-Kyei, A. P. C. Chan, *International Best Practices of Public-Private Partnership*, https://doi.org/10.1007/978-981-33-6268-0_3

3.2 Review of Past Studies on governments' Motivations/ Reasons for Adopting PPP

Since the early 1990s, there has been an increasing interest by governments globally to employ the PPP concept (Ke et al. 2009; Osei-Kyei et al. 2014). In this regard, the underlying motives for adopting PPP by governments in both developed and developing countries have been of great concern in recent times. This is because the motivation and rationale for PPP adoption underpins the establishment of performance objectives in PPP transactions, which is critical in ensuring PPP project success (Yuan et al. 2009; Liu et al. 2015). However, considering the diverse economic situations, culture and political ideologies, the reasons for PPP in the developed and developing economies/countries surely differ (Cheung et al. 2010a; Osei-Kyei et al. 2014). In the developing economies, PPP is viewed to have been adopted by governments as conditionality on loans from international financial institutions (Thomas et al. 2006; Jamali 2004). Others assert that alleviating poverty and improving the living standard of people are the major reasons for PPP in many developing countries (Bhatia and Gupta 2006).

Notwithstanding, previous empirical studies have presented some reasons for adopting PPP in developing countries. For example, Chan et al. (2009) by means of a questionnaire survey found out that the reasons for adopting PPP in China include providing an integrated solution for public infrastructure, solving the problem of public sector budget constraint and reducing public money tied up in capital investments.

Ismail (2013) also found similar reasons to that of Chan et al. (2009), where from the Malaysian perspective, PPP implementation is mainly due to its creativity and innovation approaches, solves the problem of public sector budgetary constraint and provides an integrated solution for public infrastructure.

Furthermore, Osei-Kyei et al. (2014) through a questionnaire survey also found out that PPP implementation in Ghana is chiefly due to its reduction of public administration cost, allows for shared risk and reduces the problem of public sector budget constraint. It is observed that the major motives of governments in developing countries have gone beyond relieving them from their budgetary constraint. Innovation and risk sharing offered by the PPP policy have also gain attention in developing countries. Nevertheless, the huge infrastructure deficit, which puts heavy financial pressure on budgetary allocations as well as forcing many governments to accept loans with strict conditionality, continues to be a major motivation for many governments in developing countries.

Governments in the developed countries seem to have divergent rationale for PPP adoption. Obviously, given the high infrastructure growth in these industrialized countries, the drive to adopt PPP policy is not likely to be underpinned by huge infrastructure gaps, which often put enormous pressure on government budgetary allocations as seen in developing countries. For instance, a study conducted by Liu and Wilkinson (2011) indicates that the acceleration of infrastructure provisions, better risk allocation and whole of life cycle cost savings are the key drivers for PPP in New Zealand. Also, Li et al. (2005b) summarized reasons for PPP in the U.K as better project technology and economy, greater public benefit, public sector avoidance of regulatory and financial constraints and public sector saving in transaction costs.

Cheung et al. (2010a) through a questionnaire survey emphasized that providing an integrated solution, facilitating creativity and innovative approaches and saving time in delivery project form the basic reasons for PPP implementation in Australia. Consequently, some of the reasons identified in Australia also form the rationale for PPP adoption in Hong Kong, particularly facilitating creativity, innovative approaches and providing an integrated solution for public infrastructure (Cheung and Chan 2011). This is quite understandable because in recent times, most developed countries aim to build public infrastructure incorporating green technology, where PPP is seen as the best procurement approach (Efficiency Unit 2008).

From the above pieces of literature, it is obvious that the reasons for governments accepting the PPP policy vary among developing and developed economies. Whilst, some governments primarily seek to adopt the PPP policy due to financial stress and burden given the huge infrastructure deficits; others rather embrace the policy predominantly for innovative and creativity enhancement purposes. However, there could also be some common reasons for adopting PPP by governments in both developing and developed economies/countries. Thus, it will be important for such reasons to be fully explored to further broaden knowledge on the international best practices of PPP.

3.3 Data Collection

A questionnaire survey was conducted with targeted practicing and experienced PPP practitioners in both Ghana and Hong Kong. The purpose of selecting Ghana and Hong Kong as representatives is due to the rapid growth of trade and investment (i.e. PPP investments) collaborations between Africa and Asian developed economies/countries. Thus, it is more impactful to explore the differences and similarities in the implementation practices of PPP in these diverse regions. Undoubtedly, Ghana and Hong Kong represent typical emerging and advanced economies in the African and Asian regions respectively. The questionnaire required respondents to rate on a five-point Likert scale (i.e. 1 = least important and 5 = extremely important) the importance of each reason for adopting PPP as applied in their respective jurisdiction. Targeted respondents were selected based on a two –stage sampling approach. First, a purposive sampling method with pre-defined criteria was adopted to identify initial prospective respondents. The pre-defined criteria were that: 1) respondent should have in-depth knowledge on the general practice of PPP and must have followed very closely to PPP development in Ghana or Hong Kong; and 2) respondent should have adequate direct hands-on working (at least one project) and/or research experience in PPP project delivery in Ghana or Hong Kong. In the second stage, the identified respondents were opportunistically requested to suggest potential colleagues who may be interested to participate in the research study. Majority of the suggested prospective participants willingly accepted to participate in the survey and were therefore added to the final list of respondents.

In total, 207 potential respondents from the industrial and academic sectors were sourced and identified from PPP related publications that focused on Ghana or Hong Kong, dedicated private sector organizations and public institutions/departments that have expressed strong interest in PPP projects (e.g. Ghana (Ghana Water Company Limited, Ghana Highways Authority, Ghana Ports and Harbour Authorities, Urban Roads Department, Public Procurement Authority and Public Investment Division); Hong Kong (Housing Department, Efficiency Unit, Highways department, Architectural Services Department and Civil Engineering and Development Department)). Out of the total respondents identified, 120 came from Ghana, whereas 87 came from Hong Kong. The major reason for identifying a large number of respondents in Ghana is because, in the last couple of years more public institutions and departments have initiated quite a large number of PPP projects, therefore many people are involved with PPP practice in Ghana compared to Hong Kong. Also, Ghana's population size (i.e. 26.7 million) is much bigger compared to the population size in Hong Kong (7.24 million) (World Bank Group 2015a,b, thus there is the possibility of identifying more potential PPP practitioners in Ghana than Hong Kong. Questionnaires were distributed to the targeted respondents either by face-to-face (i.e. majority of questionnaires distributed in Ghana) and/or e-mails (i.e. majority of questionnaires distributed in Hong Kong). A total of 103 completed questionnaires were received; 77 from Ghana and 26 from Hong Kong representing response rates of 64.17% and 29.89% respectively. The higher response rate in Ghana was anticipated considering that majority of the questionnaires were administered by face-to-face, which always yields a favourable response rate compared to online and telephone surveys (Ameyaw and Chan 2015a). Although, a lower response rate is obtained in Hong Kong, the sample size of 26 is considered satisfactory and reasonable when compared with past related studies that were conducted in Hong Kong (see e.g. Cheung et al. 2009, 2012a,b,c,d (34 responses); Javed 2013 (18 responses)). More importantly, the Hong Kong respondents possess rich experience in PPP practice (Table 3.1). Notwithstanding, the overall sample size of 103 is suitable and adequate for further analysis, when compared with similar studies including Cheung et al. (2009) (45 responses; 34 from Hong Kong and 11 from Australia) and Liu et al. (2016) (57 responses; 32 from China and 25 from Australia). The demographic distribution of respondents is presented in Table 3.1.

As shown in Table 3.1, approximately 80% and 77% of respondents from Ghana and Hong Kong respectively are industrial practitioners (i.e. public and private sectors), who are mostly exposed to the actualities of PPP practices compared to respondents from the academic sector. Further, almost 62% and 65% of respondents have more than six years of PPP experience either as researchers and/ or industrial practitioners in Ghana or Hong Kong. This implies that the survey respondents from both countries have enough PPP experience to offer reliable responses for analysis.

Table 3.1 Background information of survey respondents from Ghana and Hong Kong

Characteristics		Ghana No. of respondents	Ghana Percent (%)	Hong Kong No. of respondents	Hong Kong Percent (%)
Sector of PPP	Academic	15	19.50	6	23.10
	Public	35	45.50	12	46.20
	Private	27	35.10	8	30.80
	Total	77	100	26	100
Years of industrial and/or research experience	5 years and below	29	37.70	9	34.6
	6–10 years	33	42.90	6	23.1
	11–15 years	9	11.70	7	26.9
	16–20 years	5	6.50	1	3.8
	21 yrs. and above	1	1.30	3	11.5
	Total	77	100	26	100

Source: Osei-Kyei and Chan (2017) (*With permission from Emerald Group Publishing Ltd*)

3.4 Differences in Governments' Reasons/Motivations for Adopting PPP in Ghana and Hong Kong

The overall rankings of the reasons for adopting PPP in Ghana and Hong Kong is given in Table 3.2. From the table, it is noticeable that the mean values for Ghana and Hong Kong range from 2.58 to 4.48, and 2.46 to 4.69. The total variations in responses are 1.9 and 2.23 for Ghana and Hong Kong respectively. These outputs suggest that the Ghanaian respondents rated the set of reasons more similarly compared to their Hong Kong counterparts. Further, in Ghana's ranking, seven reasons emerged as critical (i.e. mean values ≥3.50) whereas in Hong Kong's ranking, nine reasons are critical. Essentially, a factor (i.e. reason) with mean value equal to or above 3.50 is considered critical because such a factor is not relatively neutral (neither important nor unimportant) but is significant and essential to practice. Overall, the Hong Kong respondents perceived the set of reasons as more relevant compared to their Ghanaian counterparts.

The significance test results on the rankings of the reasons for adopting PPP between the two independent groups are shown in Table 3.2 (see last column). The statistical test was conducted at a pre-defined significance test value of 0.05, where a factor with p-value less than 0.05 signifies significant difference on the importance of that factor among respondents from Ghana and Hong Kong. As presented in the table, five out of the twelve government reasons for adopting PPP are significantly different among the two jurisdictions. This outcome confirms previous assertion that governments in developing and developed economies/countries have different motives for adopting the PPP concept because of the varying levels of infrastructure needs (Osei-Kyei et al. 2014).

Table 3.2 Differences in the ranking of reasons for adopting PPP in Ghana and Hong Kong

	Ghana		Hong Kong		Overall		Mann-Whitney U test
Reasons	Mean	Rank	Mean	Rank	Mean	Rank	p value
Promotes quick delivery of public infrastructure projects	4.48	1	4.31	3	4.44	1	0.17
Allows for shared risks	4.22	3	4.38	2	4.26	2	0.19
Reduces government financial burden in public infrastructure provision	4.31	2	4.12	4	4.26	3	0.18
Provides reliable and quality service delivery	3.74	6	4.69	1	3.98	4	0.00*
Improves public infrastructure management and maintenance	3.87	4	3.69	6	3.83	5	0.21
Private sector's ability to generate funds	3.55	7	3.77	5	3.60	6	0.25
Facilitates creative and innovative approaches in public infrastructure development	3.44	9	3.54	9	3.50	7	0.29
Promotes technology transfer and innovation	3.77	5	2.46	12	3.44	8	0.00*
Reduces the public sector administrative costs	3.29	10	3.42	10	3.32	9	0.29
Offers benefits to local economic and social development	3.48	8	2.69	11	3.28	10	0.00*
Facilitates sustainable public infrastructure development	3.00	11	3.65	7	3.14	11	0.00*
Enhances government integrated solution capacity	2.58	12	3.58	8	2.84	12	0.00*

Source: Osei-Kyei and Chan (2018) *(With permission from VGTU Press)*

Essentially, the reasons with significant differences which are ranked higher in Ghana than Hong Kong, directly relate to the economic and social benefits of PPP. These include "promotes technology transfer and innovation", and "offers benefit to local economic and social development". The results are consistent with previous studies including Osei-Kyei et al. (2014), where economic benefits related reasons for PPP were identified to be critical in Ghana. "Promotes technology transfer and innovation" is ranked fifth by the Ghanaian respondents, whereas Hong Kong respondents ranked it 12th. Though in the traditional bid-build system, technology transfer and innovation is also relevant; it is more critical in PPPs in Ghana and other developing countries. This is because the PPP market in Ghana is dominated by foreign investors (Dulaimi et al. 2010). Apparently, PPP projects require huge investment capital and most locally based investors (i.e. small and medium enterprises) cannot enter into such project arrangement. Thus, many governments including the Government of Ghana (GoG) hope that through PPP schemes, local practitioners particularly those from the public sector will learn from the innovations and skills of these international private firms. That notwithstanding, it has been mentioned in the national policy for PPPs in Ghana that, any public project that does not seem to promote technology transfer and innovation should not be procured through PPPs (MOFEP 2011). Although the immediate effect of technology transfer may not be seen, in the long-term, it will improve the investment potentials of local investors. Also, locally based investors will be able to build their capacity to engage in more divestitures. Unlike Ghana, Hong Kong has local firms who are experienced and have the capacity to deliver PPP projects; therefore, very few foreign firms operate in their PPP markets (Li et al. 2005a). For instance, projects including the Asia World Expo, Cross Harbour Tunnel and Cyber Port project procured in Hong Kong, involved a partnership with locally based investors and companies. Certainly, the issue of technology transfer and innovation will not be part of the reasons for PPP implementation in Hong Kong as well as other developed economies/countries (Li et al. 2005a).

"Offers benefit to local economic and social development" is ranked eighth in Ghana and eleventh in Hong Kong. Though, it is ranked quite close by the two respondent groups, the difference between the mean values obtained for Ghana (i.e. 3.48) and Hong Kong (i.e. 2.69) is large. Osei-Kyei et al. (2014) explained that in general, PPP projects contribute substantially to local economic and social development through the creation of jobs and other business opportunities for local commuters. Ghana and many other developing economies/countries have high unemployment rate and poor social livelihood. Thus, many governments in developing countries anticipate that through PPP schemes the rate of unemployment could be reduced considering that PPP projects create more job opportunities compared to construction projects delivered through the traditional bid-build system. Job creation in PPP projects has been proven in some past PPP projects in developing economies/countries. For in- stance, in the Lekki toll road PPP project in Nigeria, Osei- Kyei and Chan (2015) reported that 635 short term and 1146 long term jobs were created for people. Other successful PPP projects in developing countries including the N4 toll road project in South Africa and Port of Maputo in Mozambique created many short- and long-term job opportunities for local commuters. Undoubtedly, Hong Kong has lower unemployment rate compared to Ghana; therefore, the need to adopt the PPP policy to create more job opportunities is not important to the government. Importantly, Hong Kong is an international business hub, which host large private companies from Asia and Europe. Therefore, job creation may not necessarily need to come from the implementation of PPP projects as perceived by the Ghana government.

The remaining three reasons with significant differences between the two respondent groups are ranked higher in Hong Kong than Ghana. They relate to the efficiency and quality service delivery of public facilities. They include "provides reliable and quality service delivery", "facilitates sustainable public infrastructure development" and "enhances government integrated solution capacity". These outputs support the findings of Cheung et al. (2010a,b), where efficiency related drivers were identified as the most important reasons for implementing PPP in Hong Kong and other developed economies/countries including Australia and the U.K.

The reason "provides reliable and quality service de- livery" is ranked first by respondents from Hong Kong, whereas their Ghanaian counterparts ranked it sixth. As explained by Cheung et al. (2009), Hong Kong being an international region requires high quality of public facilities and services at all time. Therefore, considering that the private sector is well-known to offer better and quality service compared to the public sector (Li et al. 2005a,b,c; Osei- Kyei et al. 2014), the PPP concept is considered as the best option to address this issue. Importantly, this is seen in the Asia World Expo project and the Cyber Port project. Emphatically, the Hong Kong government could have pro- cured the Asia World Expo project through the traditional bid-build method. However, because of the high quality of service required to keep it to an international standard (Cheung 2009); the government partnered the private sector (Hayllar 2010). Currently, it is counted as one of the world class exhibition centres and contributes greatly to Hong Kong's tourism industry (Asia World Expo 2010). Although, Ghana also requires high quality service delivery for public facilities, this is certainly not a necessity now. In the meantime, the focus of the Ghana government is to provide more public facilities to meet the increasing demand and rapid urbanization. Further, unlike Hong Kong, Ghana is not an international hub; therefore, the desire to keep public facilities at high international standards is not there. It does not form part of the priorities of the government as seen in Hong Kong and other developed countries.

"Facilitates sustainable public infrastructure development" is seventh in Hong Kong and eleventh in Ghana. This is unsurprising considering that sustainability in construction has become a very important issue in Hong Kong compared to Ghana. In the last decade, the Hong Kong government with the assistance of the Hong Kong Green Building Council (HKGBC) have made consider- able efforts to promote green buildings and infrastructure. Importantly, the PPP concept contributes substantially towards developing green infrastructure in Hong Kong. Through PPP arrangements, the government provides specific sustainability requirements, where the private sec- tor will have to consider those requirements in its design and management of the PPP project. Though, sustainability could be achieved through the traditional bid-build procurement method, it is very costly compared to the use of PPP schemes. The reason is that in PPP, the private sector can use its innovation and expertise to design, con- struct and manage the green public facility at a much reasonable cost. The Cyber Port project is a notable example of a PPP project in Hong Kong, where investor employed green technologies and environmental awareness management strategies. While, sustainability is also important in Ghana; it is an emerging concept. Hence, the government does not consider it a major priority for implementing the PPP policy. Essentially, sustainability will be an important reason for adopting PPP in Ghana, only when the basic economic and social benefits of PPP have been realised.

"Enhances government integrated solution capacity" is ranked eighth by respondents from Hong Kong, whereas their Ghanaian counterparts ranked it last (i.e. twelfth). As explained by Chan et al. (2009), PPP as an integrated solution means that all functions are bundled as a complete set of contracts and awarded. The functions include designing, financing, building, operating and maintenance. Also, small scale projects are put together as one contract and awarded to a private consortium. These enhance efficiency and reduce the total lifecycle cost of the public facility. In Hong Kong, the construction and maintenance of public infrastructure could be very expensive. Thus, the PPP concept offers the government opportunity to provide public facilities at a much-reduced cost through the bundling of functions or combining different small-scale projects as one contract, which is often not seen in the traditional procurement method. Unlike Hong Kong, providing public facilities in Ghana through the traditional method is not very expensive considering a lot of favourable factors such as the availability of land, local materials and labour force. Therefore, small-scale projects are often awarded separately to local contractors. This therefore may have contributed to the low ranking of this reason for adopting PPP in Ghana.

3.5 Similarities on the Rankings of Governments' Reasons/Motivations for Adopting PPP in Ghana and Hong Kong

The similarities in the top and bottom rankings by respondents from the two jurisdictions were analysed using quartile groupings. Table 3.3 shows the upper and lower quartiles of the reasons for adopting PPP for each country. The hinges (cut off values) for the upper quartile subsets are 4.13 for Ghana and 4.26 for Hong Kong. Further, the lower quartile hinges are 3.32 (Ghana) and 3.45 (Hong Kong). The

Table 3.3 Quartile groupings of reasons for adopting PPP in Ghana and Hong Kong

	Ghana		Hong Kong	
Quartiles	Reasons for adopting PPP	Mean	Reasons for adopting PPP	Mean
Upper quartile $(Q_3)_{Ghana}=4.13$ $(Q_3)_{HK}=4.26$	Promotes quick delivery of public infrastructure projects	4.48	Provides reliable and quality service delivery	4.69
	Reduces government financial burden in public infrastructure provision	4.31	Allows for shared risks	4.38
	Allows for shared risks	4.22	Promotes quick delivery of public infrastructure projects	4.31
Lower quartile $(Q_1)_{Ghana}=3.32$ $(Q_1)_{HK}=3.45$	Reduces the public sector administrative costs	3.29	Reduces the public sector administrative costs	3.42
	Facilitates sustainable public infrastructure development	3.00	Offers benefits to local economic and social development	2.69
	Enhances government integrated solution capacity	2.58	Promotes technology transfer and innovation	2.46

Source: Osei-Kyei and Chan (2018) *(With permission from VGTU Press)*

upper quartile subsets of both Ghana and Hong Kong contain three factors, with mean values ranging between 4.22 and 4.48, and 4.31 and 4.69 respectively.

Two reasons for adopting PPP appeared in each country's subset; these include 'promotes quick delivery of public infrastructure projects' and 'allows for shared risks'. Essentially, these reasons are also within the top three rankings of both jurisdictions. The reason 'promotes quick delivery of public infrastructure projects' is ranked first in Ghana and third in Hong Kong. Clearly, the results imply that irrespective of the culture or geographical differences, this reason for adopting PPP is critical for governments. This is quite understandable considering the fact that in PPP arrangements, the private sector gains when the project is completed before schedule. This is because projects completed on or before schedule offer investors the opportunity to recoup their returns earlier. Thus, the PPP concept tends to promote the timely delivery of public facilities. It is the best procurement option to adopt if governments want to provide public facilities at a particular period of time. The quick delivery of PPP projects is seen in many projects both in developing and developed countries. For example, the Cross-Harbour Tunnel project in Hong Kong was constructed within 36 months instead of the planned 47 months (Mak and Mo 2005). Similarly, the N4 Toll road project in South Africa was constructed within three and half years as planned. Allows for shared risk is ranked third by the Ghanaian respondents, whereas their Hong Kong counterparts ranked it second. This similarity in the top rankings of both jurisdictions is also not surprising. This is because risk management is a critical component of PPP arrangements (Ke et al. 2011). Unlike the traditional bid-build system where majority of project risks are retained by the public sector; in PPP scheme, risks are shared among parties, preferable to the party with better mitigation techniques (Chan et al. 2011). This relives the public sector from the burden of mitigating risks that are not within their capacity. The effect of risk sharing in PPP is that it reduces the lifecycle cost of the project and ensures value for money in public infrastructure delivery. In the lower quartile subsets, 'reduces the public sector administrative cost' is the only reason which falls within each jurisdiction's subset. It is ranked tenth by respondents from both countries. The public sector administration costs for public projects involve the cost of hiring consultants, project monitoring activities and mitigating more of project risks. Importantly, the public sector's administrative cost is very high for traditional bid-build projects due to the excessive risks retained by the public sector. Considering that PPP allows, the public sector to transfer more risks to the private investor, the administrative costs of the public sector are often reduced. Apparently, 'reduces public sector administrative cost' is ranked lower in both countries because of the misallocation and incomplete transfer of risks recorded in many of the recent PPP projects implemented in Ghana and Hong Kong. Because risks are not properly shared among parties or the public sector retains excessive risks than the private investor, the public sector's administrative costs end up increasing. This may have influenced the respondents to rank this reason for adopting PPP lower.

3.6 Chapter Summary

This chapter explored the reasons for governments adoption of PPP. It mainly focused on the differences among developing (Ghana) and developed economies (Hong Kong). The significant test results indicate that the Ghanaian respondents ranked the reasons related to the economic and social benefits of PPP higher whereas, the Hong Kong respondents ranked efficiency and quality services related reasons rather higher. The economic and social benefits related reasons include "promotes technology transfer and innovation", and "offers benefit to local economic and social development". Also, the efficiency and quality service-related reasons include "provides reliable and quality service delivery", "facilitates sustainable public infrastructure development" and "enhances government integrated solution capacity". Quartile grouping analysis was used to identify the similarities in the top (i.e. upper quartile) and bottom (i.e. lower quartile) rankings by respondents from both economic juris- dictions. The results show that "promotes quick delivery of public infrastructure projects" and "allows for shared risks" are the most important reasons for adopting PPP in both economic jurisdictions. This implies that irrespective of cultural and geographical differences these reasons are critical to governments. On the other hand, the reasons "reduces the public sector administrative cost" is ranked lower in both economic jurisdictions.

References

Ameyaw, E. E., & Chan, A. P. (2015a). Risk allocation in public-private partnership water supply projects in Ghana. *Construction Management and Economics, 33*(3), 187–208.

Asia World Expo. (2010). *Asia World-Expo continues to add tremendous value to Hong Kong's exhibition industry and economy*. Available from http://www.asiaworld expo.com.hk/html/en/NewsRoom/PressReleaseDetail_ff73a51fda6e4a7383c60bf523a605bf.html. Accessed on May 2015.

Bhatia, B., & Gupta, N. (2006). *Lifting constraints to public-private partnerships in South Asia*. Washington, DC: World Bank. © World Bank.

Chan, A. P. C., Lam, P. T., Chan, D. W., Cheung, E., & Ke, Y. (2009). Drivers for adopting public private partnerships—empirical comparison between China and Hong Kong special administrative region. *Journal of Construction Engineering and Management, 135*(11), 1115–1124.

Chan, A. P. C., Yeung, J. F., Calvin, C. P., Wang, S. Q., & Ke, Y. (2011). Empirical study of risk assessment and allocation of public-private partnership projects in China. *Journal of Management in Engineering, 27*(3), 136–148.

Cheung, E. (2009). *Developing a best practice framework for implementing public private partnerships (PPP) in Hong Kong* (Doctoral dissertation, Queensland University of Technology).

Cheung, E., & Chan, A. P. C. (2011). Evaluation model for assessing the suitability of public-private partnership projects. *Journal of Management in Engineering*. https://doi.org/10.1061/(ASCE)ME.1943-5479.0000044.

Cheung, E., Chan, A. P., & Kajewski, S. (2009). Reasons for implementing public private partnership projects: Perspectives from Hong Kong, Australian and British practitioners. *Journal of Property Investment & Finance, 27*(1), 81–95.

Cheung, E., Chan, P. C., & Kajewski, S. (2010a). Suitability of procuring large public works by PPP in Hong Kong. *Engineering, Construction and Architectural Management, 17*(3), 292–308.
Cheung, E., Chan, A. P., & Kajewski, S. (2010b). The public sector's perspective on procuring public works projects-comparing the views of practitioners in Hong Kong and Australia. *Journal of Civil Engineering and Management, 16*(1), 19–32.
Cheung, E., Chan, A. P. C., & Kajewski, S. (2012a). Factors contributing to successful public private partnership projects, Comparing Hong Kong with Australia and the United Kingdom. *Journal of Facilities Management, 10*(1), 45–58.
Cheung, E., Chan, A. P. C., & Kajewski, S. (2012b). Factors contributing to successful public private partnership projects, Comparing Hong Kong with Australia and the United Kingdom. *Journal of Facilities Management, 10*(1), 45–58.
Cheung, E., Chan, A. P. C., Lam, P. T. I., Chan, D. W. M., & Ke, Y. (2012c). A comparative study of critical success factors for public private partnerships (PPP) between Mainland China and the Hong Kong Special Administrative Region. *Facilities, 30*(13/14), 647–666.
Cheung, E., Chan, A. P. C., Lam, P. T. I., Chan, D. W. M., & Ke, Y. (2012d). A comparative study of critical success factors for public private partnerships (PPP) between Mainland China and the Hong Kong Special Administrative Region. *Facilities, 30*(13/14), 647–666.
Dulaimi, M. F., Alhashemi, M., Ling, F. Y. Y., & Kumaraswamy, M. (2010). The execution of public–private partnership projects in the UAE. *Construction Management and Economics, 28*(4), 393–402.
Efficiency Unit (2008). Serving the community by using the private sector - An introductory guide to Public Private Partnerships (2nd Edition), The Hong Kong Special Administrative Region, Government.
Hayllar, M. R. (2010). Public-private partnerships in Hong Kong: Good governance–The essential missing ingredient? *Australian Journal of Public Administration, 69*(1), 99–119.
Ismail, S. (2013). Positive and negative factors for the successful adoption of Public Private Partnership: evidence from Malaysia. In *Public Private Partnerships (PPP): Malaysian studies* (pp. 65–81). Kuala Lumpur: IIUM Press.
Jamali, D. (2004). Success and failure mechanisms of public private partnerships (PPPs) in developing countries, Insights from the Lebanese context. *The International Journal of Public Sector Management, 17*(5), 414–430.
Javed, A. A. (2013). *A model of output specifications for public-private partnership projects* (Doctoral dissertation, The Hong Kong Polytechnic University).
Ke, Y., Wang, S., Chan, A. P., & Cheung, E. (2009). Research trend of public-private partnership in construction journals. *Journal of Construction Engineering and Management, 135*(10), 1076–1086.
Ke, Y., Wang, S., Chan, A. P., & Cheung, E. (2011). Understanding the risks in China's PPP projects: ranking of their probability and consequence. *Engineering, Construction and Architectural Management, 18*(5), 481–496.
Li, B., Akintoye, A., Edwards, P. J., & Hardcastle, C. (2005a). Critical success factors for PPP/PFI projects in the UK construction industry. *Construction Management and Economics, 23*(5), 459–471.
Li, B., Akintoye, A., Edwards, P. J., & Hardcastle, C. (2005b). Perceptions of positive and negative factors influencing the attractiveness of PPP/PFI procurement for construction projects in the UK: Findings from a questionnaire survey. *Engineering Construction Architectural Management, 12*(125), 148.
Li, B., Akintoye, A., Edwards, P. J., & Hardcastle, C. (2005c). The allocation of risk in PPP/PFI construction projects in the UK. *International Journal of project management, 23*(1), 25–35.
Liu, T., & Wilkinson, S. (2011). Adopting innovative procurement techniques: Obstacles and drivers for adopting public private partnerships in New Zealand. *Construction Innovation: Information, Process, Management, 11*(4), 452–469.

Liu, J., Love, P., Davis, P., Smith, J., & Regan, M. (2015). Conceptual framework for the performance measurement of public-private partnerships. *Journal of Infrastructure Systems*. https://doi.org/10.1061/(ASCE)IS.1943-555X.0000210.

Liu, T., Wang, Y., & Wilkinson, S. (2016). Identifying critical factors affecting the effectiveness and efficiency of tendering processes in Public–Private Partnerships (PPPs): A comparative analysis of Australia and China. *International Journal of Project Management, 34*(4), 701–716.

Mak, C. K., & Mo, S. (2005). *Some aspects of the PPP approach to transport infrastructure development in Hong Kong*. Highways Department, Government of the Hong Kong Special Administrative Region.

Ministry of Finance and Economic Planning. (2011). *National policy on public – private partnerships*. Ghana.

Osei-Kyei, R., & Chan, A. P. C. (2015). Developing transport infrastructure in Sub-Saharan Africa through public–private partnerships: Policy practice and implications. *Transport Reviews*, 1–17.

Osei-Kyei, R., & Chan, A. P. (2017). Empirical comparison of critical success factors for public-private partnerships in developing and developed countries. *Engineering, Construction and Architectural Management.*

Osei-Kyei, R., & Chan, A. P. (2018). Comparative study of governments' reasons/motivations for adopting public-private partnership policy in developing and developed economies/countries. *International Journal of Strategic Property Management, 22*(5), 403–414.

Osei-Kyei, R., Dansoh, A., & Ofori–Kuragu, J. K. (2014). Reasons for adopting Public–Private Partnership (PPP) for construction projects in Ghana. *International Journal of Construction Management, 14*(4), 227–238.

Thomas, A. V., Kalidindi, S., & Ganesh, L. S. (2006). Modelling and assessment of critical risks in BOT road projects. *Construction Management and Economics, 24*, 407–424.

World Bank. (2015a). *Private Participation in Infrastructure (PPI), regional snapshots*. Washington, DC: World Bank. Available from: http://ppi.worldbank.org/explore/ppi_exploreRegion.aspx?regionID=1. Accessed on May 2015.

World Bank. (2015b). *Private Participation in Infrastructure (PPI), Middle East and North Africa. Regional snapshots.* Washington, DC: World Bank. Available from: http://ppi.worldbank.org/explore/ppi_exploreRegion.aspx?regionID=1. Accessed on May 2015.

Yuan, J., Zeng, A. Y., Skibniewski, M. J., & Li, Q. (2009). Selection of performance objectives and key performance Indicators in public-private partnerships projects to achieve value for money. *Construction Management and Economics, 27*(3), 253–270.

Chapter 4
Factors Contributing to the Success of Public-Private Partnership Projects

Abstract Since the evolution of PPP, few countries have succeeded in procuring many projects, whereas many others have seen little progress. Also, in the last decade, there have been several reported failed and distress PPP projects globally. Therefore, it is crucial that the best ways and practices of PPP are continuously monitored and evaluated to inform future project implementation. However, in exploring the success factors or best practices of PPP, there is a need to look at it from a cross cultural perspective rather than from a specific country/region viewpoint. In this regard, this chapter explores and discusses the success factors required for the effective implementation of PPP projects from two diverse economic regions (i.e. developing and developed economies) based on empirical data. Ghana and Hong Kong are used as representatives for developing and developed economies respectively.

Keywords Critical success factors · Best practices · Ghana · Hong Kong · Developed economies · Developing economies

4.1 Introduction

Since the evolution of PPP, few countries have succeeded in procuring many projects, whereas many others have seen little progress. Also, in the last decade, there have been several reported failed and distress PPP projects globally. Therefore, it is crucial that the best ways and practices of PPP are continuously monitored and evaluated to inform future project implementation. However, in exploring the success factors or best practices of PPP, there is a need to look at it from a cross cultural perspective rather than from a specific country/region viewpoint. In this regard, this chapter explores and discusses the success factors required for the effective implementation of PPP projects from two diverse economic regions (i.e. developing and developed economies) based on empirical data. Ghana and Hong Kong are used as representatives for developing and developed economies respectively. The questionnaire survey approach and respondents described in Chap. 3 was used to obtain data (Osei-Kyei and Chan 2018b). The PPP experts were requested to rate the importance of 32 success factors or best practices for implementing PPP projects (Osei-Kyei and Chan 2017b).

R. Osei-Kyei, A. P. C. Chan, *International Best Practices of Public-Private Partnership*, https://doi.org/10.1007/978-981-33-6268-0_4

4.2 Review of Literature on the Critical Success Factors for PPP Projects

The CSFs methodology has been widely used for management measures primarily to explore the few key areas which are necessary for a manager to achieve success (Rockart 1982). During the last couple of decades, a significant number of studies have been conducted to explore the key management techniques required for practitioners to achieve PPP projects success (Ke et al. 2009). This section will look at some key success factors which have been frequently identified by many past studies.

4.2.1 Political Stability

Political stability is one of the parameters within the political environment which is necessary to ensure a successful PPP project implementation (Jefferies 2006). Political stability is often not a great concern in the developed countries (Cheung et al. 2012a; Li et al. 2005a) however it is an important issue in developing countries particularly for those with unstable electoral system. Political stability denotes the absence of political violence and existence of a legitimate constitutional regime (Hurwitz 1973). Political stability manifest when a country observes legitimacy, orderly and peaceful way of transferring political power from one political party to the other (Iheduru 2001). Private investors are often motivated to invest in a country with adequate political stability. Moreover investors are more confident with the security of their investments in stable political environment. On the other hand, civil unrest and violence increase risks and uncertainties, which is not favourable to investors as well as the progress of PPP projects.

4.2.2 Political/Government Support

The host government's commitment and adequate consensus from opposition parties boost private investors' confidence as well as ensuring a smooth procurement process of PPP projects. It is obvious that PPP as a public policy has a direct relation with the political setting of the host country (Li et al. 2005a, b, c). Without the necessary political support, approvals for public expenditure on public projects and works would not be granted (Jacobson and Choi 2008). Moreover, the necessary support from political leaders and the host government attract more investors to a particular economy.

In jurisdictions where political backing is not strong, the political risk is considered to be high, which therefore limits competition in the tendering process, as many investors would not like to tender in such environment (Organisation for

Economic Co-operation and Development (OECD) 2008). A notable example of a country that is observed to have an overwhelming political acceptability for PPP is the U.K. (Li et al. 2005a,b,c). Despite the criticism by the Labour party when in opposition, they continued the PFI concept when assumed office in 1997. Today, the U.K. is considered the home of PFI/PPP concept with many construction projects initiated through this scheme in various sectors (Kwak et al. 2009).

4.2.3 *Good Governance*

Good governance in PPPs has substantial positive impact on the progress of PPP projects. Studies conducted by Ismail (2013a); Qiao et al. (2001); Kwak et al. (2009) indicated that measures to ensure good governance must be established to ensure PPP project success. In addition, the United Nations Economic Commission for Europe (2008) also gave similar assertion indicating good governance as an essential element which contributes to the success of PPP project implementation. Some good governance principles pointed out by UNECE (2008) include:

- Stakeholder management in PPP projects' negotiations.
- Transparency and openness with regards to decision making.
- Reporting progress of projects to the society whether good or bad.
- Efficiency in using the limited financial and human resources without any delay, waste or corruption.

Establishing a national PPP unit to oversee the procedures and process of PPP projects is also a true mark of good governance. It must be highlighted that as part of the measures to ensure good governance, responsibilities and roles must be discharged to experts and public officers with adequate knowledge in PPP procurement.

4.2.4 *Legal and Regulatory Framework*

Investors always feel secured and protected in well-structured and defined legal environment even if pockets of political unrest exist (Plummer 2013). In essence, studies conducted by Cheung et al. (2012a), Babatunde et al. (2012) and Hwang et al. (2013) pointed out a favourable legal and regulatory framework as crucial to PPP project implementation success. The existence of sound regulatory framework deepens the private investors' confidence in PPPs and more importantly, it provides enough assurance of protection against any expropriations in the partnership deal (Pongsiri 2002). Regulation allows disputes which may arise in the partnership deal to be resolved amicably by both parties (Zouggari 2003). Sound regulation encompasses a detailed tendering process, tax regime to the private investor, awarding of contract and contract management procedures (IMF 2010a). As highlighted by

Farquharson et al. (2011), questions often asked by investors concerning a country's PPP legal system when entering its PPP market include:

- The rights of the private investor
- Restrictions pertaining to the repatriation of profits by overseas investors
- Compensation mechanism when changes to a PPP arrangement arises
- Taxation mechanism on end user charges and;
- The resolution of contract disputes

4.2.5 *Coordinated Leadership and Entrepreneurship Skills*

Public officers in charge of PPP transactions need to exhibit good leadership and entrepreneurship skills because these create positive change and makes visible difference (Corrigan et al. 2005). Good leadership and entrepreneurship from PPP officers entail the abilities to coordinate the PPP arrangement process and bring all stakeholders on board to create the necessary support and commitment. Basically, good entrepreneurship and leadership does not only ensure massive stakeholder support for the project but enable the various stakeholders to share their knowledge on how the project could be successfully implemented. It is therefore necessary for public officers in charge of PPP deals to have all the required leadership and entrepreneurship qualities including integrity, vulnerability, discernment, courage, curiosity, predictability, breadth and comfort with ambiguity (Renesch 2002).

4.2.6 *Sound Economic Policy*

Economic policies have great impact on PPP projects progress. In essence, research studies conducted by Babatunde et al. (2012); Li et al. (2005a) and Zhang (2005a) pointed out the existence of good and sound economic policies as critical in boosting private investors' confidence in a country's economic system. Essentially, good economic policies have positive influence on a country's economic variables including consumer prices, inflation rate, interest rate and exchange rate (Coeure and Jacquet 2010). In certain occasions, private investors even take very risky choices in an economic system with good economic policies (Angeriz and Arestis 2007). Additionally, proper economic policies pave way for local private developers to engage in more PPP deals especially in developing countries. Good economic policies such as granting of loans with reduced interest rate to local private developers by state owned financial institutions facilitate PPP development in a country. This is because more investors would be willing to take up more projects.

4.2.7 Stable Macro-economic Indicators

Good and stable economic indicators are signs of a healthy and performing economy which favours investments. Macroeconomic indicators provide a vivid picture of the conditions and direction of an economy as well as its expansion (Griffs 2011). Macroeconomic indicators include inflation rate, exchange rates, unemployment rates, Gross Domestic Product, Gross National Product, Per Capita Income, Consumer price index etc. The stability of these indicators enables investors to assess the viability of PPP projects and more importantly, develop a good financial model (Cheung et al. 2012a; Chan et al. 2010b). On the other hand, unstable indicators make it very difficult to make realistic financial projections. However, the stability of macroeconomic conditions greatly lies in the hands of policymakers. In this regard, measures to ensure stable macroeconomic indicators need to be in place to foster PPP development and also attract more investors in a country's PPP market.

4.2.8 Mature and Available Financial Market

Highly developed, competitive and mature financial market is an incentive for investors and actually drives them to invest in jurisdictions with such markets. Essentially a mature and available financial market has been identified to have considerable impact on the success of PPP project development (Li et al. 2005a; Ozdoganm and Birgonul 2000; Jefferies 2006).

As indicated by Madura (2012), financial market is where there is an exchange of financial assets such as bonds, shares and stocks. Financial markets allow funds to be transferred from one party to another through the purchase of securities. In this regard, in a well matured financial market, investors are able to access capital from the general public for their projects (Madura 2012). In addition, a well-developed financial market enables private investors to generate funds for PPP projects at a very low financing cost (Cheung et al. 2012a). It is therefore important that governments seeking to adopt PPP schemes will develop their financial market well to allow for easy access to funds from private investors.

4.2.9 Financial Support

Given the huge investment capital required for PPP projects, it is therefore important for host governments to provide some form of financial assistance towards the project development (Tiong et al. 1992). Providing financial assistance allow private investors to raise the required amount of funds for the project. Financial assistance from governments can be in different forms depending on the host governments policy and capacity. In some countries, measures such as the Viability Gap Funding

Scheme have been put in place to provide grants in order to make PPP projects commercially viable and attractive to investors (Hodge et al. 2010). Other governments also provide financial supports in the form of senior debt subordinate, debt guarantee and tax exemptions (Cuttaree and Mandri-Perrott 2011). Nevertheless, any form of financial assistance from governments for PPP projects should be defined in the PPP policy framework, laws, guidelines and any other legal binding documents.

4.2.10 Technology Innovation

Technology innovation denotes significant technological improvements in products and process or a change in process or product (Leiringer 2003). As pointed by Jefferies et al. (2002) and Liu and Wilkinson (2013), technology innovation is essential in ensuring PPP project success because it enables the Project Company to generate enough savings from future maintenance and operational costs (Robinson and Scott 2009). It is therefore necessary for the host government to institute measures and policies to support innovations in PPP Projects and private investments in general. In some countries, regulations are introduced to protect the intellectual property of innovators like the patent right system. Essentially, such regulations grant innovators some temporary monopoly power over their innovations making the environment conducive for PPP project development. Furthermore, output specification facilitates technology innovation compared to input specification. Unlike the conventional procurement method where specific input requirements are given; in PPP arrangements, output requirements are rather given to allow private investors to employ more sophisticated innovative approaches which could help reduce future maintenance and operation costs.

4.2.11 Technology Transfer

Technology transfer involves the movement of technological know-how or knowledge among partners (individuals, enterprise and institutions) in order to enhance one's knowledge, skills and expertise (Abramson et al. 1997). Technology transfer is not often considered important to the success of PPP Projects in the developed economies/countries like the U.K (Li et al. 2005a); however, it is a significant factor to PPP projects success in the developing regions. This is because most of private developers in these countries are foreign companies (Dulaimi et al. 2010). Even in some developing countries, technology transfer is considered a requirement for initiating PPP projects (MOFEP 2011). This implies that a project which will not facilitate the transfer of technology and knowledge may not be considered through PPP approach. To ensure effective technology transfer, it is important for locally based firms to partner international firms to undertake PPP projects. In this way, local practitioners could learn any new available technologies from its foreign partner. Further, governments also have a role to play in promoting technology transfer in PPP projects. They should amend laws and regulations which restrict foreign equity holdings and licensing arrangements in order to allow free transfer of inputs by foreign companies.

4.2.12 Strong Private Consortium

Given the large and complex nature of PPP projects, it is often difficult for a single construction firm to undertake such kind of project. Thus, a Project Company often referred to as Special Purpose Vehicle (SPV) is formed to undertake PPP project. SPV allows different firms and institutions with adequate management, operational skills and capital to contribute and partake in undertaking the project. However, the structure and compatibility of this entity influences the success of PPP projects (Liu and Wilkinson 2013; Ng et al. 2012; Tang and Shen 2013). A weak and poorly managed consortium would obviously result in difficulties and eventually a failure to undertake the PPP project successfully. In this regard, consortium must be equipped with strong technical, operational and managerial capacity to be able to undertake PPP projects (Zhang 2005b). Additionally, the management board of the company must see to it that risks and benefits among stakeholders are shared accordingly. In jurisdictions where locally based companies' structures are weak particularly in developing countries, governments could assist in strengthening the local companies' capacity both financially and technically in order to put them in a favourable position to compete with international consortia for PPP projects.

4.2.13 Private Sector Financial Capabilities

The financial status and capacity of the private consortium is very essential in ensuring PPP success (Ng et al. 2012; Jefferies et al. 2002). Often private consortia raise funds for PPP projects through a mix of equity (shareholders) and debts financing (financial institutions). However, the ability of a concession consortium to raise the required amount of project funds greatly is an indication of how capable and well-structured the consortium has been formed. In this regard, it is very necessary for the management of the consortium to employ strategies to be able to raise the required amount of project funds both through the stock market and from financial institutions. On the other hand, host governments could also provide some financial support in the form of grants and debts in order to expand the consortium's financial strength.

4.2.14 Good Project Technical Feasibility

It is mostly a requisite for PPP projects to go through feasibility studies at the inception stage (OECD 2008). However, a good feasibility studies covering the economic, technical, financial and environmental issues is crucial to the success of PPP projects (Mladenovic et al. 2013; Jamali 2004; Dulaimi et al. 2010). Proper feasibility studies enable both parties (public and private sectors) to assess the practicality and viability of the proposed project. A poorly conducted feasibility study could hamper efficient and effective risk sharing mechanism which in the long run will have negative repercussions on the success of PPP projects. It must be emphasized that both the public and private sectors are responsible for conducting their feasibility studies in order to facilitate a competitive negotiation process. Public authorities could seek advisory services from the established national PPP unit when conducting detailed feasibility studies (OECD 2010).

4.2.15 Appropriate Risk Allocation and Sharing

Allocating and sharing risk is one of the fundamental components of PPP arrangement (Jin and Doloi 2008). Risk allocation involves identifying risks and appropriately sharing it among parties (public and private sectors) (Ke et al. 2010a, b). During negotiations, risks are clearly defined and allocated to the party that has better mitigation techniques to manage (Roumboutsos and Anagnostopoulos 2008). It must be highlighted that one unique feature which differentiates PPP from the conventional procurement is the allocation of risk hence a proper mechanism must be developed to help in allocating risk effectively and efficiently. It is also important for governments to refrain from the idea of transferring all project risk to the private sector as this could affect the progress or future participation of private investors in PPP projects. However, the public partner must retain risks that obviously go beyond the control of the private sector.

4.2.16 Openness and Constant Communication

The lengthy nature of PPP arrangements often requires a stable and enduring relationship among project parties. In this regard ensuring openness and frequent communication between project participants are very essential to the successful performance of PPP projects (Abdul-Aziz and Kassim 2011; Jacobson and Choi 2008; Meng et al. 2011). Openness refers to the ability of project actors to freely inquire and seek the concerns of other parties without any intimidations. Openness throughout the lifecycle of PPP process enables project parties to manage disputes which often arises (Liu et al. 2014).

4.2.17 *Clear Goals and Objectives*

Undoubtedly the host government and private consortium always have different objectives for engaging in PPP deals (Osei-Kyei et al. 2014). Nonetheless, it is always important that the key actors (public and private sectors) set a common goal and objective which offers mutual benefits. Establishing a common goal and objective helps define the roles and responsibilities of each party in the PPP arrangement (Tang et al. 2012). When each party's role is clearly defined it fosters communication and effective relationship among parties.

4.2.18 *High Level of Commitment from Both Parties*

Exhibiting a high level of commitment from both the host government and private sectors throughout the project lifecycle also facilitates the success of PPP projects (Cheung et al. 2012a; Li et al. 2005a). It must be emphasized that commitment towards a PPP project should be ensured not only during the initial stages of the project but throughout the entire lifecycle of the project. In certain occasions, a change of government could affect parties' commitment towards the project development; hence it is helpful for proper measures and contractual agreement to be adopted to protect the sanctity of the project contract. Nevertheless, a change in public authority's leadership should not necessarily affect the host government's commitment and enthusiasm towards the implementation of PPP projects.

4.3 Differences on the Rankings of Critical Success Factors (CSFs) for PPP Projects in Ghana and Hong Kong

The overall ranking of the CSFs for PPP projects is presented in Table 4.1. Specifically, the table shows the mean score analysis of CSFs for PPP projects in Ghana and Hong Kong. From the table, it is observable that the mean values of CSFs for the two jurisdictions range between 2.52 and 4.69 (Ghana) and 2.35 and 4.73 (Hong Kong). The total variations in responses are 2.17 and 2.38 for Ghana and Hong Kong respectively. Clearly, these outputs indicate that respondents from Ghana rated the CSFs more similarly than Hong Kong respondents. Further, in Ghana's ranking 18 CSFs emerged as important (i.e. mean values $\geq$3.50), whereas in Hong Kong's ranking, 16 CSFs emerged as important. This implies that respondents from Ghana generally perceived the set of CSFs as more relevant to PPP projects implementation in Ghana compared to Hong Kong respondents.

The significance test results on rankings of the CSFs for PPP projects among the two independent groups are shown in Table 4.1 (see last column). The test was conducted with the null hypothesis that no significant difference exists in the perception

Table 4.1 Differences in the ranking of the CSFs for PPPs in Ghana and Hong Kong

	Ghana		Hong Kong		Overall		Mann Whitney U test
CSFs for PPPs	Mean	Rank	Mean	Rank	Mean	Rank	P value
Favourable legal and regulatory framework	4.53	2	4.65	2	4.56	1	0.21
Transparent PPP process	4.69	1	2.96	25	4.25	2	0.00*
Clarity of roles and responsibilities among parties	4.39	4	3.69	11	4.21	3	0.00*
Political stability	4.43	3	3.58	14	4.21	4	0.00*
Appropriate risk allocation and sharing	3.70	10	4.73	1	3.96	5	0.00*
Right project identification	4.00	5	3.38	19	3.84	6	0.00*
Political/Government support	3.91	6	3.54	15	3.82	7	0.02*
Detailed project planning	3.75	9	3.92	7	3.80	8	0.26
Choosing the right private consortium	3.47	19	4.62	3	3.76	9	0.00*
High level of enthusiasm and willingness from parties	3.66	11	3.73	10	3.68	10	0.58
Clear project brief and design development	3.57	15	3.96	6	3.67	11	0.02*
Mature and available financial market	3.64	12	3.62	13	3.63	12	0.86
Reliable service delivery	3.56	16	3.77	9	3.61	13	0.25
Clear goals and mutual benefit objectives	3.39	22	4.12	5	3.57	14	0.00*
Open and frequent communication among stakeholders	3.60	14	3.50	16	3.57	15	0.52
Well organized and committed public agency/department	3.61	13	3.46	17	3.57	16	0.22
Strong private consortium	3.25	27	4.50	4	3.56	17	0.00*
Competitive tendering process	3.83	7	2.42	31	3.48	18	0.00*
Employment of competent transaction advisors	3.51	18	3.31	21	3.46	19	0.21
Reasonable user fee charge	3.31	25	3.85	8	3.45	20	0.00*
Long term demand for the project	3.38	23	3.65	12	3.45	21	0.05
Sound economic policies	3.52	17	3.23	22	3.45	22	0.07
Stable macroeconomic indicators	3.77	8	2.35	32	3.41	23	0.00*
Environmental impact of project	3.40	21	3.19	23	3.35	24	0.17
Streamline of approval process	3.34	24	3.35	20	3.34	25	0.96
Public/community support	3.13	29	3.42	18	3.20	26	0.06
Government providing guarantees	3.42	20	2.46	30	3.17	27	0.00*
Technology transfer	3.30	26	2.73	28	3.16	28	0.00*
Technological innovation	3.19	28	2.62	29	3.05	29	0.00*
Existence of a PPP project champion	2.95	31	3.12	24	2.99	30	0.25

(continued)

Table 4.1 (continued)

	Ghana		Hong Kong		Overall		Mann Whitney U test
CSFs for PPPs	Mean	Rank	Mean	Rank	Mean	Rank	P value
Public/community participation and coordination	2.97	30	2.77	27	2.92	31	0.15
Government providing financial support	2.52	32	2.81	26	2.59	32	0.06

Source: Osei-Kyei and Chan (2017b) *(With permission from Emerald Group Publishing Ltd)*

of respondents from Ghana and Hong Kong at a significance level of 0.05 (95% confidence interval). A CSF with p-value less than 0.05 rejects the null hypothesis, signifying that the Ghanaian and Hong Kong respondents view the importance of that factor differently and vice versa.

From the table, it is noticeable that out of the thirty-two CSFs, sixteen are significantly different, with p-values less than 0.05. Clearly, this confirms the uniqueness and distinct characteristics of PPP markets in Ghana and Hong Kong. Further, the results suggest that different strategic measures and procedures are required to achieve PPP project success in developing and developed countries, thus certain PPP practices in the developed countries may not equally be applicable in developing countries. The Ghanaian respondents ranked political/government support and political stability, sixth and third respectively; whereas the Hong Kong respondents ranked them 15th and 14th. Unlike, Ghana, the PPP concept has received strong political will and support from the Hong Kong government over the past two decades. Often the Hong Kong government supports and commits into certain PPP projects financially. For example, in the Hong Kong Disneyland Theme Park PPP project, the Hong Kong government invested HK$ 3.25 billion owing 57% shares in the joint venture company, whereas Disney Company invested HK$2.45 billion with 43% shares (Esty 2001). In other tunnel projects such as the Cross-Harbour Tunnel, the Hong Kong government owned 20% shares of the Project Company (Tam 1999). Aside, the financial commitments, the Hong Kong government has introduced several editions of comprehensive policy guidelines and funded several research studies on how best the PPP policy could be implemented in Hong Kong (Cheung et al. 2012a; Efficiency Unit 2008). These are clear indication of existing government supports towards the PPP concept in Hong Kong. Similarly, the political environment in Hong Kong is considerably stable, with no political violence and agitations during upcoming elections compared to Ghana. In essence, it is important for Ghana and other developing countries to adopt some of the strategies employed by the Hong Kong government towards showing commitment and support for the PPP concept.

Transparent PPP process and competitive tendering process are observed to have significant difference between the two groups of respondents. Specifically, transparent PPP process is ranked first in Ghana, whereas in Hong Kong it is ranked 25th. Also, competitive tendering process is positioned 7th and 31st by respondents from Ghana and Hong Kong respectively. This result is unsurprising because generally transparency and competition in public procurements are major concerns to most civil society groups in Ghana and other developing countries. Essentially, there is little information with respect to the financial arrangements of public construction projects including PPP projects by the general public. In the Lekki toll road project in Nigeria, Osei-Kyei and Chan (2015a) reported that most of the problems encountered in the project resulted from the lack of transparency and competition in the procurement of the project.

Unlike Ghana and other developing countries, Hong Kong ensures transparency and competition throughout the PPP process. Information regarding the financial arrangements and the project parties' responsibilities are always publicly available through the media. Notwithstanding, in Hong Kong, the Independent Commission Against Corruption (ICAC) plays a critical role in public procurement towards ensuring transparency and competition. The corruption prevention department of the ICAC monitors the progress of PPP arrangements from the initial to the completion stages, making sure that there is no conflict of interest. Ideally, Ghana and other developing countries particularly countries in sub-Saharan African region could learn from Hong Kong's practice in ensuring transparency and competition in PPP arrangements.

Other CSFs with significant differences between the two respondent groups are technology transfer and technological innovation. Although both CSFs are ranked lower by the two groups of respondents, their mean values are higher in Ghana than Hong Kong. Technology transfer and innovation are important in Ghana because the PPP market in Ghana is dominated by foreign investors. In this regard, the GoG expects that through PPPs, the skills of local practitioners would be enhanced. In Hong Kong, most PPP projects are undertaken by locally based investors except for certain special projects such as the Hong Kong Disneyland PPP project. Therefore, the transfer of technology is not a major concern in Hong Kong as well as other developed countries (Li et al. 2005a).

Stable macroeconomic indicators and government providing guarantees are also observed to have significant differences between the two respondent groups. Stable macroeconomic indicator is ranked 8th and 32nd in Ghana and Hong Kong respectively. Similarly, government providing guarantees is ranked 20th in Ghana, whereas it is ranked 30th in Hong Kong. The macroeconomic indicators which affect PPP investments include interest, inflation and exchange rates. Emphatically, these basic economic indicators have been unstable for the past five years in Ghana and other developing countries including Nigeria. The effect on PPP investments is that they make it difficult for private investors to accurately forecast their investment returns; and more importantly, it increases the overall cost of projects. An example is seen in the Lekki toll road project, where unstable exchange rate resulted in higher operational cost, which then compelled the concessionaire to increase toll fee charges (Osei- Kyei and Chan 2015a). Undeniably, achieving a stable macroeconomic condition in Ghana and other developing countries could take some period of time; hence government guarantees in PPP projects are vital. Governments in developing countries such as Ghana need to provide sovereign guarantee and other forms of guarantees to enable investors secure the required funding from financial institutions. Certainly, this would help reduce the financing cost of PPP projects. Unlike Ghana, Hong Kong has enjoyed a stable macroeconomic condition for the past two decades. One possible reason is that Hong Kong is an international business hub, which hosts the head offices of most large private organizations from Europe and China. Considering the stable macroeconomic condition, the Hong Kong government does not provide guarantees including sovereign guarantee for PPP project development. This therefore contributes to the low ranking of this CSF in Hong Kong.

The Ghanaian respondents also ranked right project identification and clarity of roles and responsibilities among parties higher than their Hong Kong counterparts. Indeed, in recent times identifying the right project for PPP procurement has become a very important issue in Ghana (PPIAF 2009). This is because public departments and agencies are not able to properly identify the appropriate public facilities, which need to be procured through PPPs. Though more public infrastructure projects are needed in Ghana and other developing countries, definitely not all public projects are suitable through PPP schemes. Therefore, to enable public departments in Ghana to identify the right projects, the government through the Ghana Investment Promotion Council (GIPC) is developing a comprehensive National Infrastructure Plan. It is hoped that this infrastructure plan would guide the various municipal and metropolitan assemblies to select the most suitable public facility for PPP schemes. Also the roles and responsibilities of parties in PPP arrangements in Ghana are sometimes confusing to the general public and the parties themselves. This is seen in the Ghana National Housing Project (GNHP), which was not implemented. In that project, the responsibilities of the government and the private investor became an issue for debate among civil society groups. This is because the government did not clearly articulate the duties of the private investor in the partnership arrangement. Moreover, the general public felt that the government was taking up more responsibilities than the private investor in the partnership arrangement. As opposed by the Hong Kong respondents, these CSFs are not critical in Hong Kong. This is understandable because proper feasibility studies and consultations are done prior to the implementation of PPP projects. This was actually seen in the Asia World Expo PPP project and Hong Kong Disneyland Theme park (Hayllar 2010).

The other CSFs which are observed to have significant differences are rather ranked higher in Hong Kong than Ghana. They include clear project brief and design development, choosing the right private consortium, reasonable user fee charges, strong private consortium, appropriate risk allocation and clear goals and mutual benefit objectives. It is noticeable that these CSFs rated higher in Hong Kong directly relates to the organisation and relationship of PPP projects rather than the socio-political and economic conditions of the project as seen in Ghana's ranking. This implies that the CSFs regarding the political and economic conditions of PPP projects are of low importance in Hong Kong compared to the ones that affect the organisation and relationship of the project. This is understandable because Hong Kong already has a good socio-political and economic environment for PPP development, so its focus is now on issues relating to the organisation of the project.

However, it must be highlighted that though the CSFs that relate to the organisation and relationship of PPP projects are ranked lower in Ghana, it doesn't mean that Ghana already observes these CSFs, and therefore they are of low relevance in Ghana. Rather, the results suggest that Ghana needs to focus first on the socio-political and economic conditions, which affect the success of PPP projects; after that, the other CSFs are likely to be more critical as seen in Hong Kong's ranking.

4.4 Similarities on the Rankings of CSFs for Implementing PPP in Ghana and Hong Kong

The similarities on the ranking of CSFs in Ghana and Hong Kong are calculated using quartile groupings and is shown in Table 4.2. Specifically, the CSFs for PPPs are grouped into upper and lower quartiles to identify the similarities in the top and bottom rankings among the two independent groups.

The upper quartile subset contains the 25% highest mean values of CSFs for PPP projects, whereas the lower quartile subset consists of 25% lowest mean values of CSFs. The cut off values (i.e. hinges) for the upper quartile subset are 3.76 and 3.83 for Ghana and Hong Kong respectively. Also, the lower quartile cut off values are 3.32 (Ghana) and 3.00 (Hong Kong). The upper quartile subsets of both Ghana and Hong Kong contain eight CSFs, with mean values ranging from 3.77 to 4.69 and 3.85 to 4.73 respectively. Interestingly, only one CSF (i.e. favourable legal and regulatory framework) appeared in both countries' upper quartile subsets; and, this CSF happens to be the only one with the same ranking position (i.e. second) between the two respondent groups.

The findings are in line with a previous study conducted by Cheung et al. (2012a), where favourable legal and regulatory framework was identified as very significant towards achieving PPP projects success in Hong Kong. Also, Abdel Aziz (2007) indicated the importance of available legal and regulatory framework in the successful implementation of PPP projects in developed countries including the U.S. Therefore, the results of this study confirm that favourable legal framework is very critical for PPP projects success irrespective of geographical and cultural differences.

Importantly, most developing countries including Ghana do not have well established regulations specifically for PPP arrangements. Although, the GoG has introduced a policy guideline for PPP practice (MOFEP 2011); certainly, this is not enough to enable legal transparency in PPP arrangements because of the high rate of political influences in public transactions. As reported by Cheung et al. (2012b), one of the key problems encountered by the Xiang-Jin Expressway located in the Hubei province of China was the poor legal structure and mechanism, which allowed officials to engage in fraudulent acts. The establishment of a well-defined PPP law would facilitate a partnership, which is free of intimidations by public officials.

Table 4.2 Similarities in the rankings of CSFs for PPP projects in Ghana and Hong Kong

	Ghana		Hong Kong	
Quartiles	CSFs for PPPs	Mean	CSFs for PPPs	Mean
Upper quartile $(Q_3)_{Ghana} = 3.76$ $(Q_3)_{HK} = 3.83$	Transparent PPP process	4.69	Appropriate risk allocation and sharing	4.73
	Favourable legal and regulatory framework	4.53	Favourable legal and regulatory framework	4.65
	Political stability	4.43	Choosing the right private consortium	4.62
	Clarity of roles and responsibilities among parties	4.39	Strong private consortium	4.50
	Right project identification	4.00	Clear goals and mutual benefit objectives	4.12
	Political/Government support	3.91	Clear project brief and design development	3.96
	Competitive tendering process	3.83	Detailed project planning	3.92
	Stable macroeconomic indicators	3.77	Reasonable user fee charge	3.85
Lower quartile $(Q_1)_{Ghana} = 3.32$ $(Q_1)_{HK} = 3.00$	Reasonable user fee charge	3.31	Transparent PPP process	2.96
	Technology transfer	3.30	Government providing financial support	2.81
	Strong private consortium	3.25	Public/community participation and coordination	2.77
	Technological innovation	3.19	Technology transfer	2.73
	Public/community support	3.13	Technological innovation	2.62
	Public/community participation and coordination	2.97	Government providing guarantees	2.46
	Existence of a PPP project champion	2.95	Competitive tendering process	2.42
	Government providing financial support	2.52	Stable macroeconomic indicators	2.35

Source: Osei-Kyei and Chan (2017b) *(With permission from Emerald Group Publishing Ltd)*

When there is a PPP law, disputes are more likely to be settled properly compared to when a policy guideline is in place.

In Hong Kong, favourable legal framework is also critical for PPP projects success. Although Hong Kong may have good existing legal mechanism for public partnerships and concessions, certain additional legal conditions are required for PPP projects success. For example, Abdel Aziz (2007) mentioned some specific conditions which need to be captured in PPP acts and regulations for developed countries; these include the authorization of the use of tolls and specific PPP modalities, limit to private sector freedom in toll settings, public evaluation of financing mechanism and the use of PPP for a specific period of time.

In the lower quartile subsets, technology transfer, technological innovation, government providing financial support, and public/community participation and coordination are the CSFs which fall within each jurisdiction's subsets. Technology transfer and technological innovation are ranked 26th and 28th by the respondents from Ghana respectively. Similarly, the Hong Kong respondents ranked them 28th and 29th respectively. This clearly shows that respondents from Ghana rated these CSFs slightly higher than the Hong Kong respondents. As previously mentioned, Ghana heavily depends on foreign investors for PPP investments. This is primarily because of the huge investment capital required for PPP arrangements. More importantly, many locally based investors do not have the capacity to compete with foreign investors. Therefore, the transfer of technology and innovation from these foreign firms to local practitioners is necessary in PPP arrangements. But that notwithstanding, the results still show that these CSFs are not among the very important CSFs in Ghana and Hong Kong, and they are of very low significance particularly in Hong Kong.

Government providing financial support is ranked 32nd and 26th by respondents from Ghana and Hong Kong respectively. It scored very low mean values by both jurisdictions. This is not very surprising because the GoG do not have the available funds to support PPP arrangements. Essentially, the Ghana government engages in PPP transactions in order to tap the private sector's capital and expertise (Osei-Kyei et al. 2014). Thus, it has very little financial commitments in PPP projects compared to the private investor. Similarly, in Hong Kong and other developed countries, governments hardly commitments themselves financially in PPP projects except for special types of PPP projects and specific reasons. This is because the developed countries have good and mature financial market, which every investor could raise substantial funds for PPP investments. Emphatically, it is always easier for private investors to raise capital for PPP investments in the developed countries such as Hong Kong compared to developing countries like Ghana.

Public/community participation and coordination is also ranked similarly by respondents from Ghana and Hong Kong (i.e. 30th (Ghana) and 27th (Hong Kong)). It scored very low mean values by both jurisdictions. The Efficiency Unit in Hong Kong has already mentioned that it is a requirement for public authorities to engage the community prior to the implementation of PPP projects (Efficiency Unit 2008). In this regard, several consultations and meetings with the general public are done by public authorities before PPP projects are implemented in Hong Kong. For example, before the implementation of the Asia World Expo PPP project in Hong Kong, series of consultations were done with the general public (Hayllar 2010). The basic reason was to make the general public feel part of the project development. This was also intended to boost demand after completion; therefore, it was unsurprising that by the end of 2009, the Asia World Expo PPP project has contributed almost HKD 9 billion to Hong Kong's economy (Asia World Expo 2010). Similarly, this CSF is of low importance in Ghana because majority of contracting authorities often engage and seek the concerns of the general public before implementing PPP projects. This is basically a recommended international best practice for PPPs.

4.5 Chapter Summary

This chapter holistically explored and discussed the critical success factors required to ensure the effective implementation of PPP projects from two diverse economies (i.e developing and developed economies). The case studies used were Ghana and Hong Kong, with each jurisdiction representing developing and developed economies respectively. The results indicated that a favourable legal and regulatory framework is very critical in both Ghana and Hong Kong. On the contrary, technology transfer, technological innovation, public/community participation and coordination and government providing financial support are ranked lower in both countries. Further analysis using the Mann-Whitney *U* test reveals that the Ghanaian and Hong Kong respondents view 16 CSFs differently. Majority of the CSFs that are ranked higher in Ghana relate to the socio-political and economic conditions of PPP projects, whereas the CSFs ranked higher in Hong Kong directly relate to the organisation and relationship of PPP projects.

References

Abdel Aziz, A. M. (2007). Successful delivery of public-private partnerships for infrastructure development. *Journal of Construction Engineering and Management, 133*(12), 918–931.

Abdul-Aziz, A. R., & Jahn Kassim, P. S. (2011). Objectives, success and failure factors of housing public–private partnerships in Malaysia. *Habitat International, 35*(1), 150–157.

Abramson, H. N., Encarnacao, J., Reid, P. P., & Schmoch, U. (Eds.). (1997). *Technology transfer systems in the United States and Germany: lessons and perspectives*. Washington, DC: National Academies Press.

Angeriz, A., & Arestis, P. (2007). Monetary policy in the UK. *Cambridge Journal of Economics, 31*(6), 863–884.

Asia World Expo. (2010). *Asia World-Expo continues to add tremendous value to Hong Kong's exhibition industry and economy*. Available from http://www.asiaworld expo.com.hk/html/en/NewsRoom/PressReleaseDetail_ff73a51fda6e4a7383c60bf523a605bf.html. Accessed on May 2015.

Babatunde, S. O., Opawole, A., & Akinsiku, O. E. (2012). Critical success factors in public-private partnership (PPP) on infrastructure delivery in Nigeria. *Journal of Facilities Management, 10*(3), 212–225.

Chan, A. P., Lam, P. T., Chan, D. W., Cheung, E., & Ke, Y. (2010b). Critical success factors for PPPs in infrastructure developments: Chinese perspective. *Journal of Construction Engineering and Management, 136*(5), 484–494.

Cheung, E., Chan, A. P. C., & Kajewski, S. (2012a). Factors contributing to successful public private partnership projects, Comparing Hong Kong with Australia and the United Kingdom. *Journal of Facilities Management, 10*(1), 45–58.

Cheung, E., Chan, A. P. C., Lam, P. T. I., Chan, D. W. M., & Ke, Y. (2012b). A comparative study of critical success factors for public private partnerships (PPP) between Mainland China and the Hong Kong Special Administrative Region. *Facilities, 30*(13/14), 647–666.

Coeuré, B., & Jacquet, P. (2010). *Economic policy: Theory and practice*. New York: Oxford University Press.

Corrigan, M. B., Hambene, J., Hudnut, W., III, Levitt, R. L., Stainback, J., Ward, R., & Witenstein, N. (2005). *Ten principles for successful public/private partnerships*. Washinton, DC: Urban Land Institute.

Cuttaree, V., & Mandri-Perrott, X. C. (2011). *Public–private partnerships in Europe and Central Asia: Designing crisis-resilient strategies and bankable projects*. Washington, DC: World Bank Publications.

Dulaimi, M. F., Alhashemi, M., Ling, F. Y. Y., & Kumaraswamy, M. (2010). The execution of public–private partnership projects in the UAE. *Construction Management and Economics, 28*(4), 393–402.

Esty, B. C. (2001). Structuring loan syndicates: A case study of the Hong Kong Disneyland project loan. *Journal of Applied Corporate Finance, 14*(3), 80–95.

Farquharson, E., de Mästle, C. T., & Yescombe, E. R. (2011). *How to engage with the private sector in public-private partnerships in emerging markets*. Washington, DC: World Bank Publications.

Griffs, M. (2011). *Economic indicators for dummies*. Hoboken: Wiley.

Hayllar, M. R. (2010). Public-private partnerships in Hong Kong: Good governance–The essential missing ingredient? *Australian Journal of Public Administration, 69*(1), 99–119.

Hodge, G. A., Greve, C., & Boardman, A. E. (Eds.). (2010). *International handbook on public-private partnership*. Northampton: Edward Elgar Publishing.

Hurwitz, L. (1973). Comparative politics. *Special Issue on Revolution and Social Change, 5*(3), 449–463.

Hwang, B. G., Zhao, X., & Gay, M. J. S. (2013). Public private partnership projects in Singapore: Factors, critical risks and preferred risk allocation from the perspective of contractors. *International Journal of Project Management, 31*(3), 424–433.

Iheduru, O. M. (Ed.). (2001). *Contending issues in African development: Advances, challenges, and the future (No. 219)*. Westport: Greenwood Publishing Group.

International Monetary Fund. (2010a). *Cameroon: Report on the observance of standards and codes— Fiscal transparency module*. Country Report No. 10/264, Washington, DC.

Ismail, S. (2013a). Critical success factors of public private partnership (PPP) implementation in Malaysia. *Asia-Pacific Journal of Business Administration, 5*(1), 6–19.

Jacobson, C., & Choi, S. O. (2008). Success factors: public works and public-private partnerships. *International Journal of Public Sector Management, 21*(6), 637–657.

Jefferies, M. (2006). Critical success factors of public private sector partnerships: A case study of the Sydney SuperDome. *Engineering, Construction and Architectural Management, 13*(5), 451–462.

Jefferies, M., Gameson, R., & Rowlinson, S. (2002). Critical success factors of the BOOT procurement system: Reflections from the stadium Australia case study. *Engineering Construction and Architectural Management, 9*(4), 352.

Jin, X. H., & Doloi, H. (2008). Interpreting risk allocation mechanism in public–private partnership projects: an empirical study in a transaction cost economics perspective. *Construction Management and Economics, 26*(7), 707–721.

Ke, Y., Wang, S., Chan, A. P., & Cheung, E. (2009). Research trend of public-private partnership in construction journals. *Journal of Construction Engineering and Management, 135*(10), 1076–1086.

Ke, Y., Wang, S., & Chan, A. P. C. (2010a). Risk allocation in public-private partnership infrastructure projects: Comparative study. *Journal of Infrastructure Systems, 16*, 343.

Ke, Y., Wang, S., Chan, A. P., & Lam, P. T. (2010b). Preferred risk allocation in China's public–private partnership (PPP) projects. *International Journal of Project Management, 28*(5), 482–492.

Kwak, Y. H., Chih, Y., & Ibbs, C. W. (2009). Towards a comprehensive understanding of public private partnerships for infrastructure development. *California Management Review, 51*(2), 51–78.

Leiringer, R. (2003). *Technological innovations in the context of public-private partnership projects*. Unpublished PhD thesis, Royal Institute of Technology, Stockholm.

Li, B., Akintoye, A., Edwards, P. J., & Hardcastle, C. (2005a). Critical success factors for PPP/PFI projects in the UK construction industry. *Construction Management and Economics, 23*(5), 459–471.
Li, B., Akintoye, A., Edwards, P. J., & Hardcastle, C. (2005b). Perceptions of positive and negative factors influencing the attractiveness of PPP/PFI procurement for construction projects in the UK: Findings from a questionnaire survey. *Engineering Construction Architectural Management, 12*(125), 148.
Li, B., Akintoye, A., Edwards, P. J., & Hardcastle, C. (2005c). The allocation of risk in PPP/PFI construction projects in the UK. *International Journal of project management, 23*(1), 25–35.
Liu, T., & Wilkinson, S. (2013). Can the pilot public-private partnerships project be applied in future urban rail development? A case study of Beijing Metro Line 4 project. *Built Environment Project and Asset Management, 3*(2), 250–263.
Liu, J., Love, P. E., Smith, J., Regan, M., & Davis, P. R. (2014). Life cycle critical success factors for public-private partnership infrastructure projects. *Journal of Management in Engineering, 31*, 04014073.
Madura, J. (2012). *Financial markets and institutions*. Boston: Cengage Learning.
Meng, X., Zhao, Q., & Shen, Q. (2011). Critical success factors for transfer-operate-transfer urban water supply projects in China. *Journal of Management in Engineering*. https://doi.org/10.1061/(ASCE)ME.1943-5479.0000058.
Ministry of Finance and Economic Planning. (2011). *National policy on public – private partnerships*. Ghana.
Mladenovic, G., Vajdic, N., Wundsch, B., & Temeljotov-Salaj, A. (2013). Use of key performance indicators for PPP transport projects to meet stakeholders' performance objectives. *Built Environment Project and Asset Management, 3*(2), 228–249.
Ng, S. T., Wong, Y. M., & Wong, J. M. (2012). Factors influencing the success of PPP at feasibility stage – A tripartite comparison study in Hong Kong. *Habitat International, 36*(4), 423–432.
Organisation for Economic Co-operation and Development. (2008). *Public-private partnerships: In pursuit of risk sharing and value for money*. Paris: OECD Publications.
Organisation for Economic Co-operation and Development. (2010). *Dedicated public-private partnership units: A survey of institutional and governance structures*. Paris: OECD Publishing.
Osei-Kyei, R., & Chan, A. P. C. (2015a). Developing transport infrastructure in Sub-Saharan Africa through public–private partnerships: Policy practice and implications. *Transport Reviews*, 1–17.
Osei-Kyei, R., & Chan, A. P. (2017b). Empirical comparison of critical success factors for public-private partnerships in developing and developed countries. *Engineering, Construction and Architectural Management*.
Osei-Kyei, R., & Chan, A. P. (2018b). Comparative study of governments' reasons/motivations for adopting public-private partnership policy in developing and developed economies/countries. *International Journal of Strategic Property Management, 22*(5), 403–414.
Osei-Kyei, R., Dansoh, A., & Ofori–Kuragu, J. K. (2014). Reasons for adopting Public–Private Partnership (PPP) for construction projects in Ghana. *International Journal of Construction Management, 14*(4), 227–238.
Ozdoganm, I. D., & Birgonul, M. T. (2000). A decision support framework for project sponsors in the planning stage of build-operate-transfer (BOT) projects. *Construction Management and Economics, 18*(3), 343–353.
Plummer, J. (Ed.). (2013). *Focusing partnerships: A sourcebook for municipal capacity building in public-private partnerships*. Routledge.
Pongsiri, N. (2002). Regulation and public private partnerships. *International Journal of Public Sector Management, 15*(6), 487–495.
PPIAF (Public-Private Infrastructure Advisory Facility). (2009). *"Unsolicited proposals." Module 5: Implementing and Monitoring*. Washington, DC: PPIAF. https://ppiaf.org/sites/ppiaf.org/files/documents/toolkits/highwaystoolkit/6/pdf-version/5-42.pdf. Mar. 2017.

Qiao, L., Wang, S. Q., Tiong, R. L. K., & Chan, T. S. (2001). Framework for critical success factors of BOT projects in China. *Journal of Project Finance, 7*(1), 53–61.
Renesch, J. (Ed.). (2002). *Leadership in a new era: Visionary approaches to the biggest crises of our time*. New York: Cosimo Inc.
Robinson, H. S., & Scott, J. (2009). Service delivery and performance monitoring in PFI/PPP projects. *Construction Management and Economics, 27*(2), 181–197.
Rockart, J. F. (1982). The changing role of the information systems executive: A critical success factors perspective. *Sloan Management Review, 24*(1), 3–13.
Roumboutsos, A., & Anagnostopoulos, K. P. (2008). Public–private partnership projects in Greece: risk ranking and preferred risk allocation. *Construction Management and Economics, 26*(7), 751–763.
Tam, C. M. (1999). Build-operate-transfer model for infrastructure developments in Asia: Reasons for successes and failures. *International Journal of Project Management, 17*(6), 377–382.
Tang, L., & Shen, Q. (2013). Factors affecting effectiveness and efficiency of analyzing stakeholders' needs at the briefing stage of public private partnership projects. *International Journal of Project Management, 31*(4), 513–521.
Tang, L., Shen, Q., Skitmore, M., & Cheng, E. W. (2012). Ranked critical factors in PPP briefings. *Journal of Management in Engineering, 29*(2), 164–171.
Tiong, R. L. K., Yeo, K. T., & McCarthy, S. C. (1992). Critical Success Factors in Winning BOT Contracts. *ASCE Journal of Construction Engineering and Management, 118*(2), 217–228.
United Nations Economic Commission for Europe. (2008). *Guide book on promoting good governance on public private partnership, UN, Geneva*. http://www.unece.org/ceci/publications/ppp.pdf. Accessed on 20 Dec 2013.
Zhang, X. Q. (2005a). Criteria for selecting the private-sector partner in public–private partnerships. *Journal of Construction Engineering and Management, 131*(6), 631–644.
Zhang, X. Q. (2005b). Critical success factors for public-private partnerships in infrastructure development. *Journal of Construction Engineering and Management, 131*(1), 3–14.
Zouggari, M. (2003). Public-private partnership: Major hindrances to the private sector's participation in the financing and management of public infrastructures via delegated management. *International Journal of Water Resources Development, 19*(2), 123–129.

Chapter 5
Risk Assessment of Public-Private Partnership Projects

Abstract Proper risk assessment is critical in PPP implementation. The risk management process and procedure in PPP arrangement are usually different from the other procurement methods particularly the traditional bid build method. This chapter explores the key risk factors associated with PPP implementation and evaluates them to identify the most critical ones. The risk assessment is discussed from developing and developed economies perspectives using Ghana and Hong Kong as case studies.

Keywords Risk factors · Risk assessment · Ghana · Hong Kong · Risk management

5.1 Introduction

Proper risk assessment is critical in PPP implementation. The risk management process and procedure in PPP arrangement are usually different from the other procurement methods particularly the traditional bid build method. This chapter explores the key risk factors associated with PPP implementation and evaluates them to identify the most critical ones. The risk assessment is discussed from developing and developed economies perspectives using Ghana and Hong Kong as case studies. The questionnaire survey approach and respondents described in Chap. 3 was used to obtain data (Osei-Kyei and Chan 2018). The PPP experts in Ghana and Hong Kong were requested to rate the probability and severity of 32 risk factors associated with PPP implementation (Osei-Kyei and Chan 2017).

5.2 Review of Risk Management Studies in PPP

The unique features and multidisciplinary stakeholders involved in PPP arrangements constitute to the complexity of PPP projects (Chan et al. 2014). Hence, PPP projects are often characterized with a plethora of risks than projects procured through the

R. Osei-Kyei, A. P. C. Chan, *International Best Practices of Public-Private Partnership*, https://doi.org/10.1007/978-981-33-6268-0_5

conventional approach (Shen et al. 2006; Chan and Ameyaw 2013). The critical assessment of PPP project risk factors and its allocation is considered very important in ensuring PPP projects success (Osei-Kyei and Chan 2015b; Li et al. 2005a).

Generally, risk management encompasses the identification, assessment, treatment and allocation of risks (Roumboutsos and Anagnostopoulos 2008; Chan et al. 2011). It has been one of the topical research areas in PPP studies over the past decades and continues to attract attention from researchers globally (Ke et al. 2009; Tang et al. 2012). A thorough risk assessment and proper risk allocation guarantees value for money and more importantly ensures good rate of return on investments (Grimsey and Lewis 2004; Ismail 2013). Thus, it is very crucial for project parties to ensure a systematic risk management throughout the PPP project life cycle (Chan et al. 2011).

The first step in conventional risk management process in PPPs is the identification of risk factors that affect PPP projects success (Ke et al. 2011; Bajaja et al. 1997). Several techniques including brainstorming, cause and effect diagram, project analogy and work breakdown structure are mostly used for risk identification. However, the widely used method of risk identification is the development of a risk checklist or register (Li et al. 2005b; Thomas et al. 2003; Ameyaw and Chan 2013; Xu et al. 2010). Many past studies have adopted a detailed risk register and used different research methods to assess the criticality of risk factors in PPP projects from different countries perspectives. For instance, Thomas et al. (2003) by means of a questionnaire survey on practitioners established eight very significant risk factors in Indian BOT projects. These include traffic revenue risk, demand risk, delay in financial closure, completion risk, cost overrun risk, debt servicing risk, delay in land acquisition risk and direct political risk.

Chan et al. (2014) through a Delphi survey and interviews identified 16 critical risk factors in PPP water infrastructure projects in China. However out of the 16 risk factors, completion risk, inflation and price change risk were found to have the most significant impact on water PPP projects in China. Also, Li et al. (2005b) from the U.K's perspective classified 46 potential risk factors in PFI projects into three levels namely the macro (exogenously), meso (endogenously) and micro (endogenously) factors. The macro level risk factors relate to the external conditions of PFI projects, the meso level relate to the bonding system of the project and the micro entails risk factors relating to stakeholder relationship. When the 46 risk factors were assessed by means of a questionnaire survey; construction time delay, construction cost overrun and operation cost overrun were identified to be the most critical risk factors in PFI projects in the U.K. (Li 2003).

Roumboutsos and Anagnostopoulos (2008) employed the risk register developed by Li et al. (2005b). Through a questionnaire survey, they established that risks related to the public sector namely project approvals and permit delays, poor public decision and making process and changes in legislation are the most significant risk factors in PPP projects developed in Greece. Further, they asserted that political risks such as unstable government and political opposition have low impact in Greece. Their outputs support the findings of Li (2003), where political risks were identified to be of low significance in the U.K. These assertions are understandable considering the enormous political acceptability for PPPs and favourable political environment in these countries. However, such assertions may be contrary in developing countries.

In Indonesia, Wibowo and Mohamed (2010) identified non availability of raw water, entry of new competitor, construction cost escalation, tariff setting uncertainty, non-availability of raw water and breach of contract by government as top significant risk factors in water PPP projects. Also, Shen et al. (2006) analysed the Hong Kong Disneyland Theme Park project in Hong Kong and established thirteen catalogues of risks. They include site acquisition risk, pollution to the land surroundings, unexpected underground conditions, risk of land reclamation, development risk, design and construction risk, changes of market conditions, inexperienced private partner, financial risk, operational risk, industrial action, policy and legal risk and force majeure.

In Turkey, Ozdoganm and Birgonul (2000) also presented six categories of risk factors in a hydro power plant project. These include market risks, financial risks, political risks, legal risks, construction risks and operational risks. Further, In Nigeria, Ibrahim et al. (2006) also adopted the risk register developed by Li et al. (2005b). By means of a questionnaire survey, they established that the endogenous factors such as unstable government, inadequate experience in PPP, availability of finance and land acquisition availability are the most significant risk factors in PPP projects.

Following the identification and assessments of risk factors, potential risks must be appropriately allocated to each party. The 'golden rule' which applies in risk allocation is that risks must be allocated to the partner with better mitigation technique (Chan et al. 2011). Nonetheless, quite a number of studies have proposed some risk allocation frameworks for PPP projects from different geographical perspectives. Ke et al. (2010) presented a risk allocation framework for Hong Kong and China. They established that political risks, legal risks, land acquisition and lack of tradition of private provision of public service should be allocated to the public sector in Hong Kong. Political and social risks, change in legislation and delays in project approvals and permits should be retain by the public sector in China. Project-specific risks factors (meso level) should be allocated to the private sector in both countries.

Li et al. (2005b) also emphasized that in the U.K's PFI, the public sector should retain site availability, political risks and relationship risks, whereas majority of project risks should be retain by the private partner. Further, they asserted that legislation changes should be shared by the public and private sectors. On the contrary, in Greece, Roumboutsos and Anagnostopoulos (2008) adopting similar risk register presented a risk allocation framework in which legislation changes were allocated to the public partner. Clearly, the above contemporary literature demonstrates the diverse risk factors among countries particularly between the developing and developed economies given the differences in culture, socio-political and economic conditions within which PPP operates. Nevertheless, there could be some common significant risk factors among developing and developed countries.

Thus, an empirical analysis of risk assessment in PPP projects in the two divergent economies will provide a better understanding on the international best practices for PPPs. Table 5.1 shows the risk categories for the identified risk factors from literature. Table 5.2 shows a summary of the risk factors in both developed and developing countries drawn from the comprehensive literature review. Table 5.3 also presents the definitions of the risk factors identified in literature.

5.3 Ranking of Risk Factors in PPP Projects in Ghana and Hong Kong

Table 5.4 shows the ranking of risk factors in PPP projects in each jurisdiction. The rankings are derived based on the risk impact values of the risk factors. The risk impact values are calculated by taking the square root of the product of probability and severity (i.e. impact = $\sqrt{}$ (probability *x* severity)) (Ameyaw and Chan 2015b). This approach of assessing risk factors has been widely used in construction management research studies.

Essentially, the top ten risk factors in Ghana's ranking could be classified as country risk factors. This is because they directly relate to the political, legal and economic conditions which affect PPP project delivery (Ameyaw and Chan 2015b). They are (descending order): corruption (impact = 4.21), inflation rate fluctuation (impact = 4.14), exchange rate fluctuation (impact = 3.96), delay in project completion (impact = 3.71), interest rate fluctuation (impact = 3.53), political

Table 5.1 Risk categories of PPP projects

Risk categories	Risk factors
Construction and Design risk	Construction changes, delay in land acquisition, Construction cost overruns, Design deficiency, delay in project completion, unavailability of labour and material
Concessionaire risk	Changes in shareholdings of the Project company, inexperienced private partner, poor quality of workmanship,
Economic risk	Exchange rate fluctuation, interest rate fluctuation, high financing cost, Inflation rate fluctuation
Legal risk	Tax regulations change, legislation changes
Market risk	Change in market demand, Tariff change
Natural risk	Force majeure, environmental risk
Operational risk	Change in technology, operational cost overruns, Project operation changes, high maintenance cost
Political risk	Corruption, political/ public opposition, poor public decision making, Political interference
Procurement risk	Project approvals and permits delays, absence of competition
Relationship risk	Conflict between partners, lack of commitment from project parties
Third party risk	Third party liabilities

Source: Osei-Kyei and Chan (2017) (*With permission from Emerald Group Publishing Ltd*)

Table 5.2 Summary of literature on the risk factors in PPP projects

Risk factors	Cheung and Chan (2011)	Ke et al. (2011)	Li (2003)	Chan et al. (2014)	Roumboutsos and Anagnostopoulos (2008)	Ameyaw and Chan (2013)	Chou et al. (2012)	Wibowo and Mohamed (2010)	Thomas et al. (2003)	Ibrahim et al. (2006)	Shen et al. (2006)	Ozdoganm and Birgonul (2000)
Corruption	x	x		x		x	x			x		
Exchange rate fluctuation	x	x		x		x				x		x
Political/public opposition					x	x			x			
Interest rate fluctuation	x	x	x	x	x					x		x
Poor public decision making	x	x		x	x		x			x		x
High financing cost	x	x	x		x	x	x				x	
Construction changes					x						x	
Change in technology		x	x									
Operational cost overruns	x	x	x	x	x	x		x		x	x	x
Change in market demand	x	x		x					x		x	x
Tariff change		x		x								x
Construction cost overruns			x		x			x	x			x
Tax regulations change			x		x							

(continued)

Table 5.2 (continued)

Risk factors	Cheung and Chan (2011)	Ke et al. (2011)	Li (2003)	Chan et al. (2014)	Roumboutsos and Anagnostopoulos (2008)	Ameyaw and Chan (2013)	Chou et al. (2012)	Wibowo and Mohamed (2010)	Thomas et al. (2003)	Ibrahim et al. (2006)	Shen et al. (2006)	Ozdoganm and Birgonul (2000)
Changes in shareholdings of the Project company		x					x					
Delay in land acquisition						x	x		x	x	x	
Inflation rate fluctuation		x	x	x	x			x				x
Project approvals and permits delays		x		x	x		x					
Project operation changes		x					x				x	
Political interference	x	x		x		x			x			
Inexperienced private partner					x						x	
High maintenance cost			x		x							
Force majeure	x							x			x	
Design deficiency			x		x	x					x	
Conflict between partners						x						
Poor quality of workmanship			x							x		

Risk factors	Cheung and Chan (2011)	Ke et al. (2011)	Li (2003)	Chan et al. (2014)	Roumboutsos and Anagnostopoulos (2008)	Ameyaw and Chan (2013)	Chou et al. (2012)	Wibowo and Mohamed (2010)	Thomas et al. (2003)	Ibrahim et al. (2006)	Shen et al. (2006)	Ozdoganm and Birgonul (2000)
Unavailability of labour and material			x							x		
Lack of commitment from project parties					x							
Third party liabilities				x								
Environmental risk											x	
Delay in project completion		x	x	x			x	x	x			
Legislation changes			x		x			x			x	x
Absence of competition		x		x								

Source: Osei-Kyei and Chan (2017) (*With permission from Emerald Group Publishing Ltd*)

Table 5.3 Definitions of risk factors in PPP projects

Risk factors	Definition
Corruption	Undue influence from both the public and private partners in the procurement process which may lead to increase in project cost
Exchange rate fluctuation	Volatility of local currency rates as against foreign currencies which may cause increase in cost of importation of equipment
Political/ public opposition	Stiff agitation from the opposition parties and public towards the project which could lead to delays and distress
Interest rate fluctuation	Unstable interest rate which could affect loans from financial institutions and also affect realistic financial projection
Poor public decision making	Risk of the public authority making wrong contractual decisions and negotiations due to the lack of PPP experience and unstandardized contract
High financing cost	The high cost of borrowing which may limit the private investors'ability to generate funds for the project or increase the total life cycle cost of project
Construction changes	Changes done during the construction process which may affect construction costs and time
Change in technology	Unexpected change in technology during operations which may affect continues delivery of project
Operational cost overruns	Unexpected increase in operational costs which is due to improper cost projections from the private sector or technology changes
Change in market demand	Fall in demand for the project service which results from high user fee charges
Tariff change	Increase, decrease or adjustment of user fees which may affect the expected rate of returns on investments
Construction cost overruns	Increase in construction cost due to improper assessments of construction cost or changes in designs
Tax regulations change	Unexpected adjustment of taxes on the returns of private sector which may affect the expected returns on investments
Change in shareholdings of the consortium	Changes in the management board of the project company which may affect the management of project service delivery
Delay in land acquisition	Unavailability of site to start construction of the project on time due to land disputes
Inflation rate fluctuation	Volatility of inflation rate which may affect financial projections of the private investor
Project approvals and permits delays	Delays in granting statutory approvals and permits for the project to start due to bureaucracy and unstructured system
Project operation changes	Changes in the operational design due to initial design deficiencies or large contract variations
Political interference	Government infers and influence on PPP project arrangements
Inexperienced private partner	Project company inability to undertake the proposed project as expected.
High maintenance cost	Increase in the cost of maintaining the project facility which could affect revenue generation expected
Force majeure	Events that are out of control by both parties and cannot be seen before contract is signed eg. Labour unrest, war etc.

(continued)

Table 5.3 (continued)

Risk factors	Definition
Design deficiency	Inadequacies in design of the facility which could cause private operators' inefficiency and construction changes
Conflict between partners	Misunderstandings and litigations over contract arrangements and variations
Poor quality of workmanship	Concessionaire employing personnel without adequate operation skills and expertise to perform adequately
Unavailability of labour and material	Shortage of labour and materials during construction which is due to the lack of proper project planning
Lack of commitment from project parties	Absence of commitment and enthusiasm from both project parties which could slow down the procurement of the project
Third party liabilities	Other external stakeholders may not properly implement their responsibility assigned to them in the contract which may affect the progress of the project
Environmental risk	Negative effects of construction and developing the project facility on the health and safety on its surroundings
Delay in project completion	Longer period of constructing the project facility due to shortage of labour, materials, improper project planning etc.
Legislation changes	Changes in the PPP policy or guidelines which may have implications on the project operation management and revenue
Absence of competition	Absence of fair and transparent tendering process, lack of competitive negotiations which could limit choosing a VFM bid

Source: Osei-Kyei and Chan (2017) (*With permission from Emerald Group Publishing Ltd*)

interference (impact = 3.48), high financing cost (impact = 3.41), construction cost overruns (impact = 3.40), political/public opposition (impact = 3.37) and project approvals and permit delays (impact = 3.32).

On the other hand, the top ranked ten risk factors in Hong Kong could be grouped as project-specific risks. This is because they relate to the design, construction, operation, relationship and organisation of PPP projects (Chan et al. 2011). They include (descending order): delay in land acquisition (impact = 4.21), operational cost overruns (impact = 4.15), construction cost overruns (impact = 3.98), delay in project completion (impact = 3.54), political interference (impact = 3.44), unavailability of labour and materials (impact = 3.44), change in market demand (impact = 3.42), high financing cost (impact = 3.31), construction changes (impact = 3.29) and design deficiency (impact = 3.23). Importantly, these outputs support the findings of previous studies including Ameyaw and Chan (2015a), where country risk factors such as corruption, foreign exchange rates, political and legal risks were identified to be critical in developing countries particularly Ghana. Further, Shen et al. (2006) identified project-specific risk factors such as site acquisition, land and surroundings, market risk, unexpected underground conditions, design and construction risks and land reclamation as significant risk factors in the Hong Kong Disneyland Theme Park project. Notwithstanding, the output is understandable considering that Ghana has very unstable political and

Table 5.4 Analysis of risk factors in PPP projects in Ghana and Hong Kong

	Ghana		Hong Kong	
Risk factors	Risk impact	Rank	Risk impact	Rank
Corruption	4.21	1	2.31	31
Exchange rate fluctuation	3.96	3	2.67	24
Political/public opposition	3.37	9	3.15	12
Interest rate fluctuation	3.53	5	2.90	20
Poor public decision making	3.02	16	2.98	18
High financing cost	3.41	7	3.31	8
Construction changes	3.08	13	3.29	9
Change in technology	2.87	20	2.55	27
Operational costs overrun	2.54	29	4.15	2
Change in market demand	3.25	11	3.42	7
Tariff change	3.03	15	2.68	23
Construction costs overrun	3.40	8	3.98	3
Tax regulations change	2.78	26	2.69	22
Changes in shareholdings of the Project company	2.55	28	2.29	32
Delay in land acquisition	3.21	12	4.21	1
Inflation rate fluctuation	4.14	2	2.56	25
Project approvals and permits delays	3.32	10	3.17	11
Project operation changes	2.81	25	3.06	15
Political interference	3.48	6	3.44	5
Inexperienced private partner	2.97	18	2.47	28
High maintenance cost	2.97	19	3.15	13
Force majeure	2.43	30	2.56	26
Design deficiency	2.82	23	3.23	10
Conflict between partners	2.86	21	3.04	16
Poor quality of workmanship	3.08	14	2.93	19
Unavailability of labour and material	2.58	27	3.44	6
Lack of commitment from project parties	2.82	24	2.99	17
Third party liabilities	2.84	22	2.45	29
Environmental risk	2.99	17	3.13	14
Delay in project completion	3.71	4	3.54	4
Legislation changes	1.92	32	2.80	21
Absence of competition	2.36	31	2.38	30

Source: Osei-Kyei and Chan (2017) (*With permission from Emerald Group Publishing Ltd*)

macroeconomic conditions and weak judicial systems, which pose threats to the successful delivery of PPP projects.

Unlike Ghana, Hong Kong and many developed countries including Australia, the U.K. and Canada, have stable political and macroeconomic conditions, more essentially, they have well- structured and efficient judicial system, which is conducive to enhance PPP project delivery.

In Ghana's ranking, the most significant risk factor is corruption. On the other hand, this risk factor is ranked 31st in Hong Kong. Corruption in PPPs denotes an undue influence in the procurement of PPP projects (Demuijnck and Ngnodjom 2011). Acts of corruption in PPPs come in different forms and they include bribery, bid shopping, unbalanced bids, bypassing competition and extorsion (Ameyaw and Chan 2015a). Chan et al. (2011) have mentioned that corruption increases the overall project cost and the risk of contract abrogation. In addition, corruption incites public and political protests and further reduces the general public's trust and confidence in PPP arrangements.

In recent years, the GoG has been battling with corrupt practices among public officials in public contract arrangements. In essence, there has been allegations of corruption in some past PPP projects attempted by the government (Chan and Ameyaw 2013). A typical example is the Ghana National Housing Project which was not implemented. In that project, opposition political parties and civil group societies alleged that the government had awarded the contract without going through a competitive bidding process. In actual fact, they asserted that the contract was awarded to a political party member (GNA 2010). This resulted in several agitations by the general public and political parties. This has also been seen in other developing countries particularly in African countries. In Nigeria, Osei-Kyei and Chan (2015a) reported that allegations of corruption in the Lekki toll road concession project led to countless demonstrations by commuters and political parties. Essentially, the major cause of corruption in Ghana and many other developing countries is the poor legal and superviosry systems. Many public officials with senior positions take advantage of the weak legal and monitory systems to influence processes and procedures (Chan et al. 2011).

Unlike Ghana, Hong Kong and other developed countries have a considerably low rate of corrupt practices. In essence, these countries have well structured and monitoring regime, which fish out acts of corruption in public contract arrangements. In Hong Kong, the ICAC has played very significant role in minimizing corruption in PPP arrangments over the past decades. The corruption prevention department of the ICAC ensures that parties from the public and private sectors strictly follow certain codes of conduct from the inception to the completion stages of the PPP project. It is therefore understandable that this risk factor is ranked very low in Hong Kong. Ideally, Ghana and other developing countries could follow Hong Kong's approach in combating corruption, this will help them to reduce the impact of corruption on PPP project delivery.

The second most significant risk factor in Ghana is inflation rate fluctuation. In Hong Kong, this risk factor is ranked 25th. This finding is in line with previous studies including Ameyaw and Chan (2015a), where inflation rate volatility was identified to be very significant in water supply PPP projects in developing countries. Notwithstanding, the output is not surprising because during the last couple of years, Ghana has experienced very high and unstable inflation rate. The consequence of inflation rate volatility is that it does not enable private investors to make realistic financial projections. Further, it causes an increase in toll charges, which may lead to user agitations and fall in demand for the public facility (Chan et al. 2011). Osei-Kyei and Chan (2015a) have explained that it may be difficult for developing countries particularly African countries to completely reduce inflation as seen in the industrialized countries, therefore governments should introduce sound economic policies that will help stabilize the inflation rate for an appreciable number of years.

With respect to Hong Kong, inflation rate is very stable compared to developing countries like Ghana. Specifically, the inflation rate in Hong Kong is averaged at 4.51% as at 2016 compared to 18.40% as at June 2016 in Ghana (Trading Economics 2016a). Clearly, inflation rate in Hong Kong is very low and more stabilized compared to that of Ghana. Nonetheless, Hong Kong is an international business hub; thus, it generally has very strong economic indicators including inflation rate and consumer price index.

The third most significant risk factor in Ghana is exchange rate fluctuation. On the other hand, respondents from Hong Kong ranked it 24th. This finding is also unsurprising because Ameyaw and Chan (2015a) has mentioned that developing countries particularly Ghana have a volatile macroeconomic condition, thus it is difficult for them to stabilize their currencies against foreign currencies. Similar to inflation rate, the volatility of foreign exchange rate affects realistic financial projections. It also increases the cost of importation particularly when most of the construction materials and equipment are imported. This was seen in the Lekki toll road project in Nigeria. Osei-Kyei and Chan (2015a) recounted that the depreciation of the Nigerian Naira currency against other foreign currencies caused an increase in the construction and operational costs. This subsequently led to toll hikes which attracted stiff opposition from commuters and political parties.

In Hong Kong, foreign exchange rate is more stable compared to that of Ghana and other developing countries. As previously mentioned, Hong Kong is an international financial center, thus their macroeconomic conditions are very stable and conducive for private investments. Importantly, governments in developing countries including Ghana may consider adopting the economic policies of Hong Kong in order to have a more stabilized foreign exchange rate.

Delay in project completion is the only risk factor in the top five significant risk factors, which is ranked similarly in both countries. Respondents from Ghana ranked it fourth; whereas their Hong Kong counterparts ranked it fourth. Essentially, delay in project completion is caused by several factors such as improper project planning, shortage of labour and compensation claims. Although, generally PPP projects are supposed to be completed on or before time compared to the traditional bid-build projects, the above-mentioned factors are likely to impede their construction. In Ghana, compensation claim is one critical factor that causes delay in completing PPP projects. This is because mostly people become displaced either of their settlements or their workplace as a result of the construction of the public facility. Certainly, these displaced people need to be compensated before construction starts. In most cases, the government does not pay compensation claims on time. This therefore results in court actions and legal battle between the people who are displaced and the project parties.

With respect to Hong Kong, the shortage of labour particularly skilled ones and underground changes cause project delays (Lo et al. 2006; Ng and Chan 2015). In essence, the labour force in the Hong Kong construction industry is ageing. As at

2013, 12% and 44% of the labour force in Hong Kong construction industry were over 60 and 50 years respectively (Ng and Chan 2015). This makes it sometimes quite difficult for project parties to engage young and new labour force to engage in large scale PPP projects. Similarly, underground changes which are caused by hard rocks barrier and underground pipes are often seen in Hong Kong. These considerably prolong the completion of public facilities particularly for underground projects.

Interest rate fluctuation is ranked as the fifth significant risk factor in Ghana. On the other hand, the Hong Kong respondents ranked it 20th. Similar to other macroeconomic indicators, interest rate volatility substantially increases the cost of financing and it also affects the accuracy in forecasting (Chan et al. 2011). In the long term, interest rate changes make it very difficult for project parties to achieve their objectives particularly for profit and reduced lifecycle cost of projects (Osei-Kyei et al. 2016). In Ghana, the interest rate has been fluctuating between 24% and 26% over the past couple of years (Trading Economics 2016b). Certainly, this is a serious threat to PPP projects success in terms of profitability and affordable user charges. It is therefore very important for the government to adopt sound economic policy measures to maintain the rate at an appreciable level. Unlike Ghana, Hong Kong's interest rate has been very stable and low for the past couple of years. It ranges between 0.5% and 0.75% (Trading Economics 2016c). Emphatically, the presence of large financial institutions in Hong Kong and sound economic measures by the government may have contributed to this level of interest rate. It is however understandable as to why this risk factor has very low probability and severity levels in Hong Kong compared to Ghana.

Among the top five significant risk factors in Hong Kong, delay in land acquisition is ranked first. On the other hand, respondents from Ghana ranked it 12th. This finding supports the outputs of Shen et al. (2006), in which site acquisition risk was identified among the significant risk factors in the Hong Kong Disneyland Theme Park project. Availability of land for construction projects is generally a problem in Hong Kong considering the limited geographical landscape (i.e. usable land is 1104 km^2). Most of developmental projects in Hong Kong are constructed on reclaimed lands. This is also not different with PPP projects. Some of the completed PPP projects developed on reclaimed lands include the Asia World Expo project, Hong Kong Disneyland Theme Park and Cyber Port project (Shen et al. 2006; Leung and Hui 2002). In most cases, civil society groups and non-governmental organizations whose objective is to protect water bodies strongly opposes land reclamation. Land reclamation becomes more debatable particularly when the project arrangement involves a partnership between the public and the private investor. Essentially, this was seen in the West Kowloon Cultural District project and the Cyber Port project (Chan et al. 2010).

Unlike Hong Kong, delay in land acquisition is not very critical in Ghana. However, it is rather the failure to compensate the people who are displaced due to land acquisition which needs attention. Emphatically, acquiring land for PPP projects is not that stressful as seen in Hong Kong. As a matter of fact, government has a lot of state lands which are readily available for private investments. Similarly, stool lands (i.e. lands belonging to chiefs) are mostly readily available for development.

The second most significant risk factor in Hong Kong is operational costs overrun. In Ghana, this risk factor is ranked 29th. Operational costs overrun arise from fall in market demand, change in technology, poor risk allocation and increase in the cost of operational equipment and materials. In Hong Kong, the major cause of operational cost overruns is the fall in market demand for public facility/services and improper risk allocation. Essentially, these were seen in the Western Harbour Crossing project. In that project, the demand for the project fell drastically because of high toll charges and existence of other competing projects such as the Cross-Harbour Tunnel project (Tam 1999). This has made the investor incurred excess costs overrun in operating the project.

With respect to Ghana, operational costs overrun is considered low risk primarily because few projects are in operation and majority are at the preparatory stages (Osei-Kyei and Chan 2016). Thus, Ghana does not have much operational experience in PPP projects compared with Hong Kong, where most projects are currently in operation. Certainly, with time this risk factor may become very significant when more projects reach the operational stage. The third most significant risk factor in PPP projects in Hong Kong is construction costs overrun. In Ghana, this risk factor is ranked eight. Similarly, to operational costs overrun, this risk factor is caused by increase in construction materials and equipment, changes in design and changes in ground conditions. In Hong Kong, construction cost could be very high because of unforeseen underground ground changes and land acquisition. Changes in designs due to rock barriers and underground condition may require additional specialist and experts which therefore result in unplanned construction costs. Also, land reclamation is very capital intensive. It therefore has the possibility of increasing construction costs excessively considering the experts and special equipment required. In Ghana, this risk factor is not as critical in Hong Kong because generally, construction projects are not expensive as seen in Hong Kong. More importantly, the nature and scope of PPP projects in Ghana are not in large scale form compared to PPP projects in Hong Kong. For instance, the complexity and capital required of a market shed project and a water treatment plant in Ghana are not the same as underground water tunnel projects and world class exhibition centre in Hong Kong. Importantly, the conditions and complexities of PPP projects in Ghana are not likely to cause cost overruns in the construction stage.

Political interference is ranked the fifth most significant risk factors in PPP projects in Hong Kong. In Ghana, this risk factor is ranked sixth. It is observable that this risk factor has quite similar likelihood and severity in both countries. Clearly, it has very high impact on PPP projects success in both countries. Political interference causes delays and more importantly, raises doubts among the general public on the transparency and accountability of the project.

In Hong Kong, political interference is often seen in PPP projects in the form lengthy debate by the Legislative Council (LegCo) members on PPP arrangements. In most cases, a section of the LegCo may prefer the traditional bid-build procurement system for public facilities because of the availability of public funds for such projects. Emphatically, this was seen in the West Kowloon Cultural District project. In that project, Chan et al. (2010) reported that the suitability of the project through the PPP scheme became a subject for debate among the LegCo members for a very long time. The debate caused the delay of the project for many years.

This is also similar in Ghana. However, in the Ghanaian context the debate is not often related to the suitability of the public facility through PPPs but the transparency and accountability of the tendering process. This was seen in the Ghana National Housing project. Political parties were strongly opposed to the project because of lack of competition and fairness in the awarding of the contract. This subsequently led to the failure of the project as it was not implemented.

Aside using tables, another traditional method for evaluting risk factors is to plot Risk Probability-Impact Matrix. This method of assessing risks provides a clear spectrum of the probability and severity of risk factors (Chan et al. 2011). Figures 5.1 and 5.2 show the risk impact matrix for PPP projects in Ghana and Hong Kong respectively.

The probability of the risk factors is shown on the y-axis and the severity of risk factors is on the x-axis. The scale on both axis range from 1 to 5, where 1-low critical and 5 – most critical. However, the impact matrix is divided into two categories. Risk factors with probability and severity values between 3.00 and 2.00 and those with probability values ranging between 2.00 and 3.00 but have severity values above 3.00 are considered as moderate risks. Further, risk factors with probability and severity values above 3.00 are calssified as high risks.

As shown in Figs. 5.1 and 5.2; in Ghana fourteen risk factors, which account for 44% of all the thirty-two risk factors are within the moderate category. On the other hand, eighteen risk factors, accounting for 56% of the total risk factors are considered moderate risks in Hong Kong. In the high risk category, eighteen are for Ghana and fourteen belong to Hong Kong. Further, observation of the two risk impact matrices

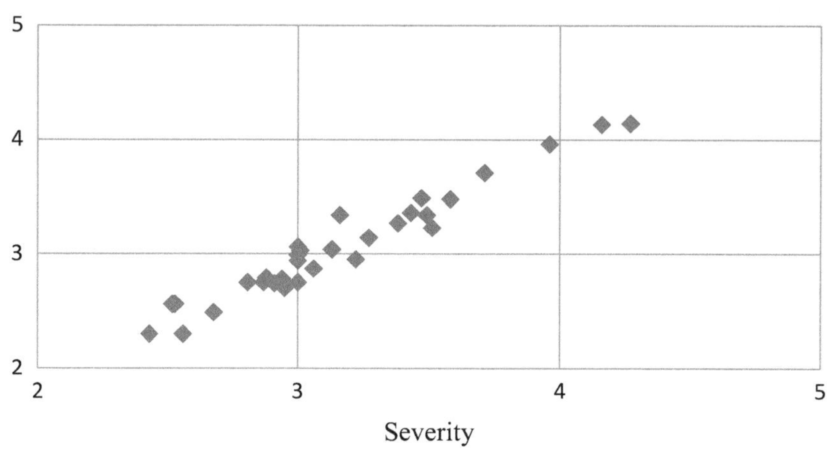

Fig. 5.1 Risk impact matrix for PPP projects in Ghana. (Source: Osei-Kyei and Chan (2017). *With permission from Emerald Group Publishing Ltd*)

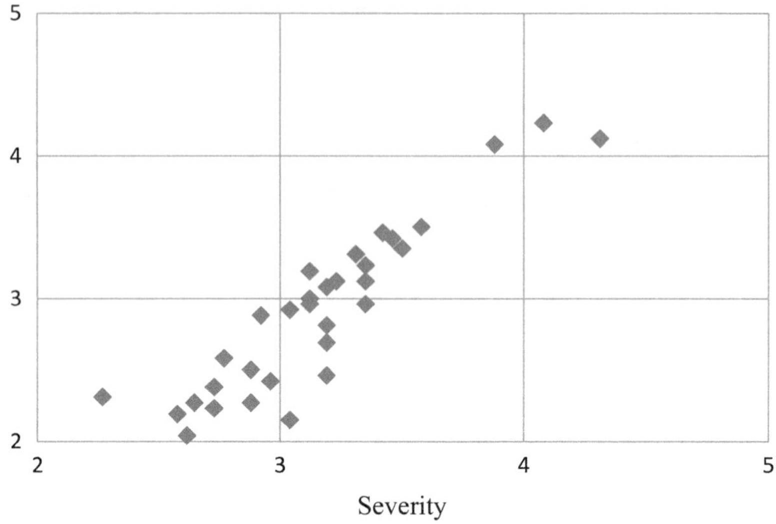

Fig. 5.2 Risk impact matrix for PPP projects in Hong Kong. (Source: Osei-Kyei and Chan (2017). *With permission from Emerald Group Publishing Ltd*)

indicate that the pattern of risk factors in Ghana is quite steeper than the pattern of risk factors in Hong Kong. This reaffirms the fact that in general, majority of the risk factors have high impact on the successful delivery of PPP projects in Ghana compared to Hong Kong. This is in line with outputs of Ameyaw and Chan (2015a) and Chan and Ameyaw (2013). They explained that considering the unfavourable economic, social and political conditions in developing countries, the success of PPP projects are likely to be affected by many risks. Thus, investors should properly plan and adopt robust strategic measures before engaging in PPP arragements in developing countries particularly Ghana and other sub-Saharan African countries.

5.4 Chapter Summary

This chapter explored the critical risk factors in PPP projects in developing and developed countries using Ghana and Hong Kong as examples. Results from the mean rankings show that respondents from Ghana ranked country risk factors more significant, whereas their Hong Kong counterparts ranked project-specific risk factors as more significant. The country risk factors relate to the political, social, legal and economic conditions which affect the success of PPP projects, whereas the project-specific risk factors relate to the design, construction, operation, relationship and organisation of PPP projects. The top five significant risk factors in Ghana are corruption, inflation rate fluctuation, exchange rate fluctuation, delay in project completion and interest rate fluctuation. In Hong Kong, the top five significant risk factors are delay in land acquisition, operational cost overruns, construction cost overruns, delay in project completion and political interference. The findings of this chapter inform international investors of the management strategies to adopt when entering into PPP arrangements in any part of the world, that is whether developing or developed countries. Also, governments who are yet to adopt the PPP concept will be considerably informed of the potential obstacles that can affect the success of PPP implementation.

References

Ameyaw, E. E., & Chan, A. P. C. (2013). Identifying public-private partnership (PPP) risks in managing water supply projects in Ghana. *Journal of Facilities Management, 11*(2), 152–182.

Ameyaw, E. E., & Chan, A. P. (2015a). Evaluation and ranking of risk factors in public–private partnership water supply projects in developing countries using fuzzy synthetic evaluation approach. *Expert Systems with Applications, 42*(12), 5102–5116.

Ameyaw, E. E., & Chan, A. P. C. (2015b). Risk ranking and analysis in PPP water supply infrastructure projects: an international survey of industry experts. *Facilities, 33*(7/8), 428–453.

Bajaja, D., Oluwoye, J., & Lenard, D. (1997). An Analysis of Contractor's approaches to Risk Identification in New South Wales, Australia. *Construction Management and Economics, 15*, 363–369.

Chan, A. P. C., & Ameyaw, E. E. (2013). "The private sector's involvement in the water industry of Ghana", *Journal of Engineering. Design and Technology, 11*(3), 251–275.

Chan, A. P., Lam, P. T., Chan, D. W., Cheung, E., & Ke, Y. (2010). Potential obstacles to successful implementation of public-private partnerships in Beijing and the Hong Kong special administrative region. *Journal of Management in Engineering*. https://doi.org/10.1061/(ASCE)0742-597X(2010)26:1(30).

Chan, A. P. C., Yeung, J. F., Calvin, C. P., Wang, S. Q., & Ke, Y. (2011). Empirical study of risk assessment and allocation of public-private partnership projects in China. *Journal of Management in Engineering, 27*(3), 136–148.

Chan, A. P., Lam, P. T., Wen, Y., Ameyaw, E. E., Wang, S., & Ke, Y. (2014). Cross-sectional analysis of critical risk factors for PPP water projects in China. *Journal of Infrastructure Systems, 21*(1), 04014031.

Cheung, E., & Chan, A. (2011). Risk factors of public-private partnership projects in China: Comparison between the water, power, and transportation sectors. *Journal of Urban Planning and Development, 137*(4), 409–415.

Chou, J. S., Tserng, H. P., Lin, C., & Yeh, C. P. (2012). Critical factors and risk allocation for PPP policy: Comparison between HSR and general infrastructure projects. *Transport Policy, 22*, 36–48.

Demuijnck, G., & Ngnodjom, H. (2011). Public-private partnerships and corruption in developing countries: A case study. *Business & Professional Ethics Journal – Center for Applied Philosophy, 3*, 253–269.

Ghana News Agency. (2010). *Issues criticized in STX housing deal not addressed – Minority*. Available from: http://www.ghananewsagency.org/print/18673. Accessed on 5 Nov 2014.

Grimsey, D., & Lewis, M. K. (2004). *Public and private partnerships. The worldwide revolution in infrastructure provision and project finance* (p. 01060-3815). Northampton: Edward Elgar Publishers.

Ibrahim, A. D., Price, A. D. F., & Dainty, A. R. J. (2006). The analysis and allocation of risks in public-private partnerships in infrastructure projects in Nigeria. *Journal of Financial Management of Property and Construction, 11*(3), 149–163.

Ismail, S. (2013). Positive and negative factors for the successful adoption of Public Private Partnership: evidence from Malaysia. In *Public Private Partnerships (PPP): Malaysian studies* (pp. 65–81). Kuala Lumpur: IIUM Press.

Ke, Y., Wang, S., Chan, A. P., & Cheung, E. (2009). Research trend of public-private partnership in construction journals. *Journal of Construction Engineering and Management, 135*(10), 1076–1086.

Ke, Y., Wang, S., & Chan, A. P. C. (2010). Risk allocation in public-private partnership infrastructure projects: Comparative study. *Journal of Infrastructure Systems, 16*, 343.

Ke, Y., Wang, S., Chan, A. P., & Cheung, E. (2011). Understanding the risks in China's PPP projects: ranking of their probability and consequence. *Engineering, Construction and Architectural Management, 18*(5), 481–496.

Leung, B. Y., & Hui, E. C. (2002). *Option pricing for real estate development: Hong Kong Disneyland. Journal of Property Investment & Finance, 20*(6), 473–495.

Li, B. (2003). *Risk management of construction public private partnership projects*. Ph.D. thesis, Glasgow Caledonian University, UK.

Li, B., Akintoye, A., Edwards, P. J., & Hardcastle, C. (2005a). Critical success factors for PPP/PFI projects in the UK construction industry. *Construction Management and Economics, 23*(5), 459–471.

Li, B., Akintoye, A., Edwards, P. J., & Hardcastle, C. (2005b). The allocation of risk in PPP/PFI construction projects in the UK. *International Journal of project management, 23*(1), 25–35.

Lo, T. Y., Fung, I. W., & Tung, K. C. (2006). Construction delays in Hong Kong civil engineering projects. *Journal of Construction Engineering and Management, 132*(6), 636–649.

Ng, J. Y. K., & Chan, A. H. S. (2015). *The Ageing Construction Workforce in Hong Kong: A Review*. Proceedings of the International Multi-Conference of Engineers and Computer Scientists 2015 Vol II, IMECS 2015, March 18–20, 2015, Hong Kong.

Osei-Kyei, R., & Chan, A. P. C. (2015a). Developing transport infrastructure in Sub-Saharan Africa through public–private partnerships: Policy practice and implications. *Transport Reviews*, 1–17.

Osei-Kyei, R., & Chan, A. P. C. (2015b). Review of studies on the critical success factors for public–private partnership (PPP) projects from 1990 to 2013. *International Journal of Project Management, 33*(6), 1335–1346.

Osei-Kyei, R., & Chan, A. P. C. (2016). Implementing public–private partnership (PPP) policy for public construction projects in Ghana: critical success factors and policy implications. *International Journal of Construction Management*. https://doi.org/10.1080/15623599.2016.1207865.

Osei-Kyei, R., & Chan, A. P. (2017). Risk assessment in public-private partnership infrastructure projects: empirical comparison between Ghana and Hong Kong. *Construction Innovation*, 204–223.

Osei-Kyei, R., & Chan, A. P. (2018). Comparative study of governments' reasons/motivations for adopting public-private partnership policy in developing and developed economies/countries. *International Journal of Strategic Property Management, 22*(5), 403–414.

Osei-Kyei, R., Chan, A. P. C., Javed, A. A., & Ameyaw, E. E. (2016). Critical success criteria for public-private partnership projects: International experts' opinion. *International Journal of Strategic Property Management, 21*.

Ozdoganm, I. D., & Birgonul, M. T. (2000). A decision support framework for project sponsors in the planning stage of build-operate-transfer (BOT) projects. *Construction Management and Economics, 18*(3), 343–353.

Roumboutsos, A., & Anagnostopoulos, K. P. (2008). Public–private partnership projects in Greece: risk ranking and preferred risk allocation. *Construction Management and Economics, 26*(7), 751–763.

Shen, L. Y., Platten, A., & Deng, X. P. (2006). Role of public private partnerships to manage risks in public sector projects in Hong Kong. *International Journal of Project Management, 24*(7), 587–594.

Tam, C. M. (1999). Build-operate-transfer model for infrastructure developments in Asia: Reasons for successes and failures. *International Journal of Project Management, 17*(6), 377–382.
Tang, L., Shen, Q., Skitmore, M., & Cheng, E. W. (2012). Ranked critical factors in PPP briefings. *Journal of Management in Engineering, 29*(2), 164–171.
Thomas, A. V., Kalidindi, S. N., & Ananthanaraynan, K. (2003). Risk perception analysis of BOT road project participants in India. *Construction Management and Economics, 21*(4), 393–407.
Trading Economics. (2016a). *Hong Kong Inflation rate: Data from 1981–2016*. Available from: http://www.tradingeconomics.com/hong-kong/inflation-cpi. Accessed on 29 July 2016.
Trading Economics. (2016b). *Ghana Interest rate: Data from 2002–2016*. Available from: http://www.tradingeconomics.com/ghana/interest-rate. Accessed on 29 July 2016.
Trading Economics. (2016c). *Hong Kong Interest rate: Data from 1998–2016*. Available from: http://www.tradingeconomics.com/hong-kong/interest-rate. Accessed on 29 July 2016.
Wibowo, A., & Mohamed, S. (2010). Risk criticality and allocation in privatised water supply projects in Indonesia. *International Journal of Project Management, 28*(5), 504–513.
Xu, Y., Chan, A. P. C., & Yeung, J. F. (2010). Developing a fuzzy risk allocation model for PPP projects in China. *Journal of Construction Engineering and Management, 136*(8), 894–903.

Chapter 6
Model for Assessing the Success Index of Public-Private Partnership Projects

Abstract Achieving PPP project success is paramount to project parties and stakeholders. However, the success of PPP projects has remained abstract which does not help in the effective implementation of the PPP policy. Therefore, it is important that a pragmatic approach is adopted to put the success of PPP projects in quantifiable terms to generate accurate and reliable information about the performance of PPP projects. Further, it is important that the success of different PPP projects should be compared to inform future implementation. This chapter develops a pragmatic tool that can be used to quantify the success levels of PPP projects.

Keywords Success model · Fuzzy Synthetic Evaluation · Success criteria · Project success index · Factor analysis

6.1 Introduction

Achieving PPP project success is paramount to project parties and stakeholders. However, the success of PPP projects has remained abstract which does not help in the effective implementation of the PPP policy. Therefore, it is important that a pragmatic approach is adopted to put the success of PPP projects in quantifiable terms to generate accurate and reliable information about the performance of PPP projects. Further, it is important that the success of different PPP projects should be compared to inform future implementation. This chapter develops a pragmatic tool that can be used to quantify the success levels of PPP projects. The questionnaire survey approach and respondents for Ghana described in Chap. 3 was used to obtain data (Osei-Kyei and Chan 2018b). The model developed in this chapter will enable practitioners to determine the success levels of their projects in a more practical and objective manner. Additionally, the evaluation model makes it possible to compare the success levels of different PPP projects on the same basis for benchmarking purposes.

R. Osei-Kyei, A. P. C. Chan, *International Best Practices of Public-Private Partnership*, https://doi.org/10.1007/978-981-33-6268-0_6

6.2 Review of Success Criteria for PPP Projects

Project success has been defined in diverse ways by researchers and each definition varies depending on the type and size of project (Parfitt and Sanvido 1993; Chan and Chan 2004). Generally, project success is described as the achievement of some externally perceived criteria (Ashley et al. 1987; Rosenau 1984). Whereas criteria denote a principle or standard by which something is judge (Lim and Mohamed 1999). Essentially, the criteria for assessing construction projects success have been debated for the past decades and there exists no acceptable checklist (Brown and Adams 2000; Ahadzie et al. 2008). Nevertheless, project success is traditionally assessed based on three major criteria namely: time; cost and quality (Cheung et al. 2000; Khosravi and Afshari 2011). These success criteria are what Atkinson (1999) referred to as the 'iron triangle'.

Over time, researchers have criticized the exclusive use of the conventional success measures of time, cost and quality for construction projects. Some have suggested the incorporation of other subjective success measures. For instance, Ahadzie et al. (2008) identified environmental impact and customer satisfaction as important additional success measurements aside the traditional set of success criteria for mass housing building projects (MHBP) in developing countries. Furthermore, Toor and Ogunlana (2010) emphasized that safety, effectiveness, satisfaction of stakeholders, efficient use of resources and reduced conflicts are the success criteria which are very important in assessing large scale construction projects success compared to the traditional measures. Also, Westerveld (2003) strongly opined that clients' appreciation, contracting partners' appreciation, stakeholders' appreciation, users' appreciation and project personnel appreciation are important success measures aside the conventional criteria of time, cost and quality. Emphatically, the traditional success criteria do not fully incorporate the proper implementation of construction projects but mainly focus on the contribution of profit (Cserháti and Szabó 2014). In addition, the traditional measures are geared towards the satisfaction of project clients and contractors without considering other external stakeholders' expectations (Westerveld 2003; Al-Tmeemy et al. 2011). In this regard, it is always essential for project managers to establish a clear set of success criteria which integrate both objective and subjective measures in order to properly evaluate construction projects success (Cox et al. 2003).

Though there is a large strand of literature on success criteria, importantly, majority of past studies have focused on the traditionally procured projects. Considering the complexity and value of PPP projects, clearly, just a hand full of success criteria of traditionally procured projects may be applicable to PPP projects. However, there are strands of literature that have attempted to explore the performance measures; performance indicators and performance objectives of PPPs from which a well-defined set of success criteria for PPP projects could be developed. Dixon et al. (2005) through semi-structured interviews with 11 PFI stakeholders found out that different groups of stakeholders have different ways of measuring PFI projects success. They pointed out that feedback from users, efficient and cost-effective procurement process, time, budget and meeting specifications are some of the key success criteria for PFI projects. Also, Zhang (2006a) examined the public client's best value-objectives in PPPs based on a questionnaire survey of international PPP experts. He grouped the public client's best value-objectives into five major components; these include: 'exploring private finance initiatives for enhanced infrastructure development'; 'maximizing the benefits to the public sector'; 'improving construction engineering and management process'; 'utilizing private sector technologies and management skills for innovation and improved efficiencies'; and 'improving the scope for private sector participation to promote the development of priority and other needed projects'.

Yuan et al. (2009) identified and evaluated 15 performance objectives of PPP projects based on a comprehensive literature review and questionnaire survey of experts. They found that acceptable quality of project; quality public service and project executed within budget are the top three performance objectives of PPP projects. Further, they identified 48 performance indicators of PPP projects and grouped them into five major components. These include: physical characteristics of projects indicators; innovation and learning indicators; stakeholders indicators; project process indicators; and financial and marketing indicators. Also, Mladenovic et al. (2013) through brainstorming with experts along with a thorough review of literature presented a two-layer approach for measuring the performance of PPP projects. They indicated that the first layer should focus on the fulfilment of the ultimate objectives of the stakeholders, whilst the second layer examines the fulfilment of the individual performance objectives of each stakeholder (i.e. public, private and users). The ultimate objectives identified include customer and owner satisfaction for the private sector, profitability, efficiency, value for money, environmental impact of project, level of services and effectiveness.

Furthermore, Liu et al. (2015) through an extensive review of literature developed a life-cycle conceptual model for measuring the performance of PPP projects. The model consists of five principal components of performance measures. These include: stakeholder satisfaction; strategies; processes; capabilities; and stakeholder contribution. Liyanage and Villalba-Romero (2015) also analysed toll road projects and developed a matrix of 11 key performance indicators and 32 performance measures which were categorized from the perspectives of project management, stakeholder and contract management.

Drawing on the above strands of literature that focused on the success criteria for general construction projects and PPP related projects, a concise and comprehensive list of fifteen PPP projects success criteria as perceived by the three major stakeholders to PPP projects is deduced. Table 6.1 shows the definition of each criterion and respective literature sources.

Table 6.1 Success criteria for PPP projects

Success Criteria	Descriptions	Sources												
		1	2	3	4	5	6	7	8	9	10	11	12	13
Profitability	A continuous income/ profit is received by parties during project operation		X									X		
Long-term relationship and partnership	Cordial relationship and well-established coordination are instituted among stakeholders													X
Satisfying the need for public facility/service	An implemented PPP project satisfies fully the need for a public facility/service												X	X
Adherence to time	Project is constructed on/ before time schedule for commissioning								X					
Adherence to budget	Project is constructed according to the estimated cost and it is without any operational cost overruns					X			X	X				
Reduced litigations and disputes	Contract litigations and disputes are minimized throughout the project lifecycle		X		X									
Reduced public sector administrative cost	Lower cost is incurred by the public sector in the administration of the project because major project risks are allocated to the private sector							X			X			X
Effective technology transfer and innovation	Technical knowledge and innovation are effectively shared among stakeholders particularly with local practitioners											X		
Local economic development	Project contributes to the economic development of the community within which the project is developed				X									
Environmental performance	Project does not affect the health and safety of occupants or the environment				X		X	X						
Reduced project life cycle cost	Lower life cycle cost is realised, which enhances the project's value for money													X

(continued)

Table 6.1 (continued)

		Sources												
Success Criteria	Descriptions	1	2	3	4	5	6	7	8	9	10	11	12	13
Reliable and quality service operations	Continuous and uninterrupted project services are provided and according to the satisfaction of users			X					X					
Meeting output specifications	Project meets the expected output standards/ requirements and delivery	X							X	X				
Effective risk management	Risks are properly identified. The risk sharing and transfer mechanism are agreed and effectively implemented by the public and private parties										X		X	
Reduced public and political protests	The reduction of agitations and protests which often arise due to increases in tariffs, lack of transparency, corruption etc.							X						

1= Lam and Javed (2015), 2= Chan and Chan (2004), 3= Meng et al. (2011), 4 = Zhang (2006a), 5= Cheung et al. (2000), 6=Chan et al. (2002), 7= Zhang (2006b), 8= Dixon et al. (2005), 9= Liyanage and Villalba-Romero (2015), 10= Li et al. (2005c), 11= Mladenovic et al. (2013), 12 = Liu et al. (2015), 13 = Yuan et al. (2009)
Source: Osei-Kyei and Chan (2017d) *(With permission from ASCE)*

6.3 Developing a Project Success Index (PSI) for PPP Projects

The data analysis involves four major steps: (1) selecting the critical success criteria (CSC); (2) identifying the critical success criteria groupings (CSCG) for PPP projects; (3) generating a PSI for each CSCG for PPP projects; and (4) developing a PSI model for PPP projects.

6.3.1 Selection of Critical Success Criteria for PPP Projects

Table 6.2 shows the mean score analysis and ranking of success criteria for PPP projects. It is noticeable that the mean values range from 2.70 (satisfying the need for public facility) to 4.40 (profitability).

Therefore, in selecting the critical success criteria, only factors with normalized values equal to or greater than 0.50 are considered (i.e. mean values ≥ 3.50). This selection mechanism has been used by many previous studies to determine the most significant factors (Xu et al. 2010b; Ameyaw et al. 2015; Chan et al. 2014). As presented in the table (see column 5), nine success criteria emerged as critical. These

Table 6.2 Ranking of success criteria for PPP projects

PPP projects success criteria	Mean	Std. Deviation	Rank	Normalization[a]
Profitability	4.40	0.80	1	1.00
Meeting output specifications	4.32	0.62	2	0.96
Adherence to budget	4.23	0.65	3	0.90
Adherence to time	4.14	0.72	4	0.85
Reliable and quality service operations	3.87	0.61	5	0.69
Effective risk management	3.78	0.77	6	0.63
Environmental performance	3.64	0.76	7	0.55
Local economic development	3.56	0.75	8	0.50
Reduced litigations and disputes	3.55	0.82	9	0.50
Reduced public and political protests	3.49	0.88	10	0.47
Long-term relationship and partnership	3.38	0.89	11	0.40
Effective technology transfer and innovation	3.36	0.90	12	0.39
Reduced public sector administrative cost	2.96	0.72	13	0.15
Reduced project life cycle cost	2.84	0.80	14	0.08
Satisfying the need for public facility/service	2.70	0.97	15	0.00

[a]Normalized value= (actual value – minimum value)/(maximum value –minimum value)
Osei-Kyei and Chan (2017d) *(With permission from ASCE)*

include (descending order): profitability (CSC1), meeting output specifications (CSC2), adherence to budget (CSC3), adherence to time (CSC4), reliable and quality service operations (CSC5), effective risk management (CSC6), environmental performance (CSC7), local economic development (CSC8) and reduced litigations and disputes (CSC9). The selected critical success criteria are then used in the next stage of analysis.

6.3.2 *Identification of Critical Success Criteria Groupings (CSCGs) for PPP Projects*

The Factor Analysis (FA) technique is used to identify CSCGs based on the mean scores of the nine critical success criteria. Generally, before conducting FA, a variable to sample size ratio of 1:5 is recommended (Lingard and Rowlinson 2006). Therefore, considering that the variable to sample ratio for this study is 1:8.5 (i.e nine variables and sample size of 77); the FA technique is considered appropriate and suitable for the data set. Notwithstanding, several preliminary tests including reliability (internal consistency), correlation matrix, Kaiser-Meyer-Olkin (KMO) and Bartlett's test of Sphericity are conducted to further ascertain the appropriateness of the technique in this study (Norusis 2008).

First, reliability test is conducted using the Cronbach's alpha model. The overall alpha value for the nine CSC is 0.761, which is above the recommended threshold of 0.70 (Chan et al. 2010b). This therefore indicates a good uniformity among the survey responses (Cheung and Chan 2011b). The correlation matrix measures the relationship among factors based on the partial correlation coefficients. The matrix calculated indicates a strong correlation among the nine CSC because their correlation coefficients exceed 0.30 for more than one other variable (Li et al. 2005a; Norusis 2008). The KMO statistic and Bartlett's test of Sphericity measure the sampling adequacy. The KMO statistic values vary between 0 and 1, where a value of 0.50 is considered acceptable for a satisfactory factor analysis (Norusis 2008). A KMO value of 0.769 is recorded, which exceeds the acceptable threshold (Field 2013). This again signifies the appropriateness of the survey data for factor analysis (Chan et al. 2010b). Further, the value of Bartlett's test of Sphericity is large (chi-square = 133.653) and its associated significance value is less than 0.05 (p = 0.00), indicating that the population correlation matrix is not an identity matrix (Norusis 2008).

The principal factor extraction (with varimax rotation) produced three-factor solution, with eigenvalues greater than 1.0, explaining 58.68% of the variance in responses (Table 10.2). Varimax rotation was chosen because it simplifies interpretation as the principal component factor is represented by small number of variables which are interpretable (Osei-Kyei et al. 2014). As shown in Table 6.3, all factor loadings of variables are close or above 0.50, with seven of them exceeding 0.60. Again, this reaffirms the appropriateness of the survey data for FA (Ahadzie et al. 2008).

Table 6.3 CSCGs for PPP projects and their underlying CSC

No.	Success criteria	Factor loadings	Eigenvalue	% of variance explained	Cumulative % of variance explained
CSCG 1	**Local development and disputes reduction**		**3.127**	**34.747**	**34.747**
CSC 8	Local economic development	0.704			
CSC 7	Environmental performance	0.618			
CSC 9	Reduced litigations and disputes	0.616			
CSC 5	Reliable and quality service operations	0.603			
CSC 6	Effective risk management	0.588			
CSCG 2	**Profit**		**1.144**	**12.717**	**47.463**
CSC 1	Profitability	0.855			
CSC 4	Adherence to time	0.760			
CSCG 3	**Cost and technical specifications**		**1.009**	**11.217**	**58.68**
CSC 3	Adherence to budget	0.819			
CSC 2	Meeting output specifications	0.486			

Osei-Kyei and Chan (2017d) *(With permission from ASCE)*

The three derived CSCGs (factor groupings) are labelled as (As shown in Table 10.2):

CSCG 1 – Local development and disputes reduction
CSCG 2 – Profit
CSCG 3 – Cost and technical specifications

It must be highlighted that the suggested label for each CSCG is subjective; hence other researchers may label it differently (Chan et al. 2010b). Nevertheless, the label for a particular CSCG is informed by variable(s) with high factor loadings (Ahadzie et al. 2008).

6.3.3 *Generating a PSI for Each CSCGs for PPP Projects*

The FSE is the main tool used to develop a PSI for each CSCG, which then enabled the development of a PSI for PPP projects. Before conducting the fuzzy modelling, two different levels are established. The first level is the three CSCGs, whereas the second level is the nine CSC. The fuzzy evaluation model is therefore demonstrated as in subsequent sections.

6.3.3.1 Calculate the Appropriate Weightings of CSC (Second Level) and CSCGs (First Level)

The weightings of CSC and CSCGs are computed based on the mean values obtained from the survey (see Table 6.4). For instance, from Table 10.3, given that CSCG2- Profit consists of two CSC with a total mean value of 8.54. The weighting of, for example, profitability (CSC1) is given as:

$$W_{CSC1} = \frac{4.40}{4.40 + 4.14} = \frac{4.40}{8.54} = 0.515$$

Therefore, the weightings of all CSC and CSCGs for PPP projects are calculated using the same procedure.

6.3.3.2 Determine the Membership Functions for CSC and CSCGs

The membership functions are derived from levels 2 to 1. This implies that the membership functions of CSC are obtained first before calculating the membership functions for each CSCG. Membership function for a CSC is obtained from the ratings furnished by the practitioners given the grades for selection (i.e. L_1-least important to L_5- extremely important).

Table 6.4 Weightings for CSC and three CSCGs for PPP projects

No.	Factors	MS for CSC	Weightings for each CSC	Total MS for each CSCG	Weightings for each CSCG
CSC 8	Local economic development	3.56	0.193		
CSC 7	Environmental performance	3.64	0.198		
CSC 9	Reduced litigations and disputes	3.55	0.193		
CSC 5	Reliable and quality service operations	3.87	0.210		
CSC 6	Effective risk management	3.78	0.205		
CSCG 1	**Local development and disputes reduction**			**18.4**	**0.518**
CSC 1	Profitability	4.40	0.515		
CSC 4	Adherence to time	4.14	0.485		
CSCG 2	**Profit**			**8.54**	**0.241**
CSC 3	Adherence to budget	4.23	0.495		
CSC 2	Meeting output specifications	4.32	0.505		
CSCG 3	**Cost and technical specifications**			**8.55**	**0.241**
	Total mean for CSCGs			35.49	

Osei-Kyei and Chan (2017d) *(With permission from ASCE)*

For example, 11.7% of the respondents rated 'adherence to budget' (CSC3) as important, 53.2% as very important and 35.1% as extremely important.

Therefore, the membership function for CSC3 is given as follows:

$$MF_{CSC3} = \frac{0.00}{L_1} + \frac{0.00}{L_2} + \frac{0.12}{L_3} + \frac{0.53}{L_4} + \frac{0.3\ 5}{L_5}$$

This function is also expressed as (0.00, 0.00, 0.12, 0.53, 0.35). Using the same approach, the membership functions for the nine CSC can be obtained (see Table 6.5).

Determining the membership functions of CSC at level 2 set the basis to calculate the membership functions for each CSCG at level 1.

$D = W_i \circ R_{i,}$ where Wi is the weightings of all CSC within each CSCG and Ri is the fuzzy evaluation matrix.

Using 'Cost and technical specifications' (CSCG 3) as an example, its membership function is calculated as:

Table 6.5 Membership functions for all CSC and CSCGs for PPP projects

No.	CSC and CSCGs	Weightings for CSC	Membership Functions at level 1 (CSC)					Membership Function at Level 2 (CSCGs)				
Local development and dispute reduction												
CSC 8	Local economic development	0.193	0.01	0.04	0.40	0.47	0.08	0.01	0.04	0.35	0.49	0.12
CSC 7	Environmental performance	0.198	0.00	0.04	0.42	0.42	0.13					
CSC 9	Reduced litigations and disputes	0.193	0.01	0.08	0.35	0.47	0.09					
CSC 5	Reliable and quality service operations	0.210	0.00	0.01	0.22	0.65	0.12					
CSC 6	Effective risk management	0.205	0.00	0.03	0.35	0.44	0.18					
Profit												
CSC 1	Profitability	0.515	0.00	0.03	0.12	0.29	0.57	0.00	0.02	0.14	0.39	0.45
CSC 4	Adherence to time	0.485	0.00	0.01	0.16	0.51	0.33					
Cost and technical specifications												
CSC 3	Adherence to budget	0.495	0.00	0.00	0.12	0.53	0.35	0.00	0.00	0.10	0.53	0.38
CSC 2	Meeting output specifications	0.505	0.00	0.00	0.08	0.52	0.40					

Osei-Kyei and Chan (2017d) *(With permission from ASCE)*

$$D_{CSCG3} = (0.495,\ 0.505) \times \begin{vmatrix} 0.00 & 0.00 & 0.12 & 0.53 & 0.35 \\ 0.00 & 0.00 & 0.08 & 0.52 & 0.40 \end{vmatrix}$$
$$= (0.00,\ 0.00,\ 0.10,\ 0.53,\ 0.38)$$

In similar manner, the membership functions of the remaining two CSCGs are derived, this is shown in Table 6.5 (column 5).

After computing the membership functions at Level 1, the project success index for each CSCG is calculated. For instance, the PSI for 'Cost and technical specifications' is computed as:

$$PSI_{(CSCG3)} = (0.00,\ 0.00,\ 0.10,\ 0.53,\ 0.38) \times (1,2,3,4,5) = 4.32$$

Using the same approach, the PSI for the other two CSCGs are determined as follows (As shown in Table 6.6):

Table 6.6 PSI for CSCGs for PPP projects

No.	CSCGs	Project Success Index (PSI)	Coefficients[a]
CSCG 1	Local development and disputes reduction	3.70	0.301
CSCG 2	Profit	4.27	0.347
CSCG 3	Cost and technical specifications	4.32	0.352
	Total	12.29	1.00

[a]Coefficient = (PSI for CSCG / $\sum$PSI for CSCG)
Osei-Kyei and Chan (2017d) *(With permission from ASCE)*

$$PSI_{(CSCG1)} = (0.01, 0.04, 0.35, 0.49, 0.12) \times (1,2,3,4,5) = 3.70$$
$$PSI_{(CSCG2)} = (0.00, 0.02, 0.14, 0.39, 0.45) \times (1,2,3,4,5) = 4.27$$

6.3.4 Developing Overall PSI for PPP Projects

A linear and additive model is used to develop a composite indicator for evaluating PPP projects success. A linear model is chosen because the three CSCGs for PPP projects are not correlated with each other. This implies that they are non-linear. As explained by Yeung et al. (2009), a linear but not non-linear model is deemed appropriate if the correlation among variables is insignificant. Therefore, given that the correlation coefficients among the CSCGs are zero, a linear and additive approach is considered suitable. Nonetheless, previous studies including Hu et al. (2016) and Lam et al. (2008) developed a performance index using an additive and linear model.

However, before developing the composite indicator, the PSIs for all CSCGs which function as coefficients in the linear model are further normalized so that they sum to one or fit within unity (see Table 8). It is logical and valid to normalize the PSIs because it provides a better reflection of the relative activity between each variable in the linear equation. Most essentially, it allows different scale of measurement for CSC to be used in the evaluation model. The project success index for PPP projects is therefore expressed by the following equation:

$$\begin{aligned} PSI = {} & (0.301 \times \text{local development and disputes reduction}) + (0.347 \times \text{profit}) \\ & + (0.352 \times \text{cost and technical specifications}) \end{aligned}$$

From the equation above, it is noticeable that cost and technical specifications obtained the highest coefficient (0.352) in the evaluation model, followed by profit (0.347) and local development and disputes reduction (0.301). The sum of all coefficients is one, which fits within unity. Essentially, the success index equation developed would considerable enable practitioners to evaluate the success level of their PPP projects in a more practical and reliable manner. Moreover, the evaluation model makes it possible for practitioners to compare the success levels of two or more projects on the same basis. This will improve the implementation practices of PPPs.

6.3.4.1 Cost and Technical Specifications (CSCG 3)

This CSCG accounts for 11.217% of the total variance in the factor analysis. It has a PSI of 4.32 and a coefficient value of 0.35 in the evaluation model. Essentially, cost and technical specifications have also been mentioned by previous studies as critical criteria for most traditionally procured projects (Chan and Chan 2004; Al-Tmeemy et al. 2011). This implies that meeting the estimated cost and technical specifications will remain a critical success criterion irrespective of the procurement method adopted for construction projects (Lam et al. (2008). This category of CSC consists of two sub factors, these include adherence to budget and meeting output specifications.

Adherence to budget has the highest factor loading (0.819) within this group. Globally, it is proven that PPP projects are very expensive to procure and could require huge investment capital (Li et al. 2005b; Cheung et al. 2009; Boussabaine 2013). However, PPP projects procured are generally more expensive and could end up in cost overruns considering the plethora of risks characterised with such projects and the high private sector borrowing in most countries (Ameyaw and Chan 2015b; World Bank 2014e). In this regard, meeting the expected cost which includes both construction and operational costs is considered very essential towards achieving success (Babatunde et al. 2012). Importantly, adopting technical innovation and good feasibility studies are some critical measures needed to ensure the adherence to budget (Jamali 2004; Dulaimi et al. 2010; Abdul-Aziz and Kassim 2011).

Meeting output specification is the second CSC within this category. It has a factor loading of 0.486. Output specifications directly relates to technical specifications of construction projects. In fact, some researchers refer output or technical specifications as quality (Chan et al. 2002; Chan and Chan 2004). Adhering to output standards is a basic requirement in PPP project implementation (Lam and Javed 2015). Output based requirements are mostly used in PPPs compared to the input-based specification often seen in the traditional design-bid-build procurement method. Notwithstanding, there still remain the fact that private developers should strictly comply with the output requirements when executing PPP projects. However, for this criterion to be achieved, the public sector must roll out very clear and concise output requirements at the tendering stage of the PPP process (Askar and Gab-Allah 2002).

6.3.4.2 Profit (CSCG 2)

Profit accounts for 12.717% of the variance explained in the factor analysis, with a PSI of 4.27. Its coefficient value in the evaluation model is 0.347. This CSCG consist of two CSC; profitability and adherence to time; each with a factor loading of

0.855 and 0.760 respectively. Undoubtedly, the sub factors directly relate to the profit received by parties from the PPP project.

Profitability is the highest loading factor in this category, and this is unsurprising. Essentially, World Bank (2009) has stressed that PPP projects in developing countries (i.e. African countries) that are not likely to generate adequate profit to parties particularly the private investor stand the chance of not attracting potential foreign investors and will not be successful. However, it is worth noting that profitability may not be a critical criterion for PPP projects implemented in most developed countries like Hong Kong. This is because most developed countries consider other critical criteria such as effective risk management and safety compared to profit (Osei-Kyei et al. 2016). Importantly, for profitability to be achieved in developing countries, critical factors including stable macroeconomic conditions, good feasibility studies, long-term demand for project and reasonable user fees need to observe by practitioners (Chan et al. 2010b; Babatunde et al. 2012; Ozdoganm and Birgonul 2000; Ismail 2013a).

Basically, time has both positive and negative repercussions on profit or investment returns in PPP project implementation. In essence, if a project is constructed before schedule, the parties are offered the opportunity to recoup their investment returns earlier and vice versa. Time performance is often affected by delays due to political debates and public agitations over PPP contracts, the right of way and compensation claims (Babatunde et al. 2015; Osei-Kyei and Chan 2015a). Even, in some situations, a PPP project could reach financial close, but the private investor cannot execute the project due to one or two of the above reasons. Thus, transparency and competition, proper stakeholder management, community and political support are some of the very crucial measures to ensure time performance.

6.3.4.3 Local Development and Disputes Reduction (CSCG 1)

This CSCG accounts for 34.747% of the total variance of the factor analysis. It has a PSI of 3.70, and its associated coefficient value in the evaluation model is 0.301. The factor consists of five CSC, where all the criteria relate to local development and the minimization of disputes in PPP project implementation. The sub factors include local economic development, environmental performance, reduced litigations and disputes, reliable and quality service operations and effective risk management.

Local economic development is the highest factor loading within this category. Ideally, it is one of the success criteria peculiar to developing countries. It refers to the economic benefits including employment opportunities that are associated with the implementation of PPP projects (Osei-Kyei et al. 2016). Generally, most governments in developing countries battle with high unemployment rates and lack of proper public facilities. Hence, they have embraced the PPP policy with the hope of creating more job opportunities and better livelihood in the local communities (Bhatia and Gupta 2006; Osei-Kyei et al. 2014). Importantly, right project identification, transparency, competition, good feasibility studies and public participation are some measures to ensure the achievement of this success criterion in Ghana and other developing countries.

Environmental performance has the second highest factor loading in this group (0.618) and it has a direct relation with local development. Basically, because PPP projects are complex and large-scale in nature, their environmental friendliness is very important to practitioners (Zhang 2006a). The development of a PPP project should not negatively affect the health and safety of the local population and its surroundings. In most developing countries, it is a requirement for environmental impact assessments of proposed PPP projects to be conducted prior to their implementation (MOFEP 2011). This signifies that though developmental projects (i.e. PPP projects) are essential for economic growth and poverty alleviation, the health and safety of the public and environment cannot be undermined. Ideally, thorough and extensive environmental assessments, public participation and integration are required to achieve this criterion.

Reduced litigations and disputes is the third CSC within this group and it has loading coefficient of 0.616. Given the lengthy nature of PPP contract arrangements, dispute is bound to occur. However, its magnitude could be minimized when due diligence is observed between parties. Very often, disputes in PPP project arrangements result in contract abrogation, high unplanned costs and damage to reputation and relationship (Cisse et al. 2013); thus, it is essential to devise proper regulatory regime which could help minimize disputes over PPP contracts. Notwithstanding, critical factors including openness, transparency, clear roles and responsibilities and clear mutual objectives are important in minimizing disputes. Reliable and quality service operation has a factor loading of 0.603. This criterion implies that the private operator provides continuous and reliable services to the satisfaction of end users/public (Osei-Kyei et al. 2016). This CSC directly relates to local development because improved and quality public facilities/services contribute to the better livelihood of the local population. Essentially, reliable and quality service operation can be achieved, if measures including consistent project monitoring, timely rectification of operational problems and public integration are strictly observed.

Effective risk management is the last CSC within this group. This success criterion refers to the proper identification, equitable allocation and effective treatment of risks (i.e. project and country risks) among project parties (Osei-Kyei et al. 2016). Undoubtedly, effective risk management has a relation with disputes minimization. Apart from the poor contracting management of public institutions; improper risk assessments and incomplete transfer of risks cause disputes over PPP contracts in most African countries. Some public contracting authorities either take up excessive risks or fail to completely transfer risks to the private partner. In the long run, this results to disputes over the treatment of risks by the project parties. Therefore, it is very important for practitioners to be conscious of risk management when implementing successful PPP projects. Ideally, measures including good feasibility studies, appropriate risk allocation and sharing, proper project planning and well-organized public department can ensure effective risk management.

6.4 Chapter Summary

PPP projects success is the ultimate goal of practitioners in both developing and developed countries. However, it is problematic to quantify PPP projects success considering that success means different things to different stakeholders. Based on data obtained from a developing country (i.e., Ghana) (Osei-Kyei and Chan 2017a), a model for evaluating PPP projects success is presented using fuzzy set theory. The research findings show that success criteria for PPP projects can be grouped into three major categories. These include Local development and disputes reduction (CSCG 1), Profit (CSCG 2), and Cost and technical specifications (CSCG 3). Cost and technical specification is the most significant critical success criteria grouping with an index of 4.32, followed by profit (4.27) and local development and disputes reduction (3.70). By obtaining the project success index for each success criteria grouping, a linear and additive model is adopted to develop a pragmatic PSI model for PPP projects. The research output provides an opportunity for practitioners to objectively and comprehensively determine the success level of their PPP projects.

References

Abdul-Aziz, A. R., & Jahn Kassim, P. S. (2011). Objectives, success and failure factors of housing public–private partnerships in Malaysia. *Habitat International, 35*(1), 150–157.

Ahadzie, D. K., Proverbs, D. G., & Olomolaiye, P. O. (2008). Critical success criteria for mass house building projects in developing countries. *International Journal of Project Management, 26*, 675–687.

Al-Tmeemy, S. M. H. M., Abdul-Rahman, H., & Harun, Z. (2011). Future criteria for success of building projects in Malaysia. *International Journal of Project Management, 29*(3), 337–348.

Ameyaw, E. E., & Chan, A. P. (2015b). Evaluation and ranking of risk factors in public–private partnership water supply projects in developing countries using fuzzy synthetic evaluation approach. *Expert Systems with Applications, 42*(12), 5102–5116.

Ameyaw, E. E., Chan, A. P., Owusu-Manu, D. G., & Coleman, E. (2015). A fuzzy model for evaluating risk impacts on variability between contract sum and final account in government-funded construction projects. *Journal of Facilities Management, 13*(1), 45–69.

Ashley, D. B., Lurie, C. S., & Jaselskis, E. J. (1987). Determinants of construction project success. *Project Management Journal, 18*(2), 69–79.

Askar, M., & Gab-Allah, A. (2002). Problems facing parties involved in build operate, and transport projects in Egypt. *Journal of Management in Engineering, 18*(4), 173–178.

Atkinson, R. (1999). Project management: cost, time and quality, two best guesses and a phenomenon, it's time to accept other success criteria. *International Journal of Project Management, 17*(6), 337–342.

Babatunde, S. O., Opawole, A., & Akinsiku, O. E. (2012). Critical success factors in public-private partnership (PPP) on infrastructure delivery in Nigeria. *Journal of Facilities Management, 10*(3), 212–225.

Babatunde, S. O., Perera, S., Zhou, L., & Udeaja, C. (2015). Barriers to public private partnership projects in developing countries: A case of Nigeria. *Engineering, Construction and Architectural Management, 22*(6), 669–691.

Bhatia, B., & Gupta, N. (2006). *Lifting constraints to public-private partnerships in South Asia.* Washington, DC: World Bank. © World Bank.

Boussabaine, A. (2013). *Cost planning of PFI and PPP building projects*. London: Routledge.

Brown, A., & Adams, J. (2000). Measuring the effect of project management on construction outputs: a new approach. *International Journal of Project Management, 18*, 327–335.

Chan, A. P. C., & Chan, A. P. (2004). Key performance indicators for measuring construction success. *Benchmarking: An International Journal, 11*(2), 203–221.

Chan, A. P., Scott, D., & Lam, E. W. (2002). Framework of success criteria for design/build projects. *Journal of Management in Engineering*. https://doi.org/10.1061/(ASCE)0742-597X(2002)18:3(120).

Chan, A. P., Lam, P. T., Chan, D. W., Cheung, E., & Ke, Y. (2010b). Critical success factors for PPPs in infrastructure developments: Chinese perspective. *Journal of Construction Engineering and Management, 136*(5), 484–494.

Chan, A. P., Lam, P. T., Wen, Y., Ameyaw, E. E., Wang, S., & Ke, Y. (2014). Cross-sectional analysis of critical risk factors for PPP water projects in China. *Journal of Infrastructure Systems, 21*(1), 04014031.

Cheung, E., & Chan, A. (2011b). Risk factors of public-private partnership projects in China: Comparison between the water, power, and transportation sectors. *Journal of Urban Planning and Development, 137*(4), 409–415.

Cheung, S. O., Tam, C. M., Ndekugri, I., & Harris, F. C. (2000). Factors affecting client's project dispute resolution satisfaction in Hong Kong. *Construction Management Economic, 18*(3), 281–294.

Cheung, E., Chan, A. P., & Kajewski, S. (2009). Reasons for implementing public private partnership projects: Perspectives from Hong Kong, Australian and British practitioners. *Journal of Property Investment & Finance, 27*(1), 81–95.

Cisse, H., Menon, N. M., Segger, M. C. C., & Nmehielle, V. O. (2013). *The World Bank legal review. Volume 5: Fostering development through opportunity, inclusion, and equity*. Washington, DC: World Bank Publications.

Cox, R. F., Issa, R. R. A., & Ahrens, D. (2003). Management's perception of key performance indicators for construction. *Journal of Construction Engineering and Management, 129*(2), 142–151.

Cserháti, G., & Szabó, L. (2014). The relationship between success criteria and success factors in organisational event projects. *International Journal of Project Management, 32*(4), 613–624.

Dixon, T., Pottinger, G., & Jordan, A. (2005). Lessons from the private finance initiative in the UK: Benefits, problems and critical success factors. *Journal of Property Investment & Finance, 23*(5), 412–423.

Dulaimi, M. F., Alhashemi, M., Ling, F. Y. Y., & Kumaraswamy, M. (2010). The execution of public–private partnership projects in the UAE. *Construction Management and Economics, 28*(4), 393–402.

Field, A. P. (2013). *Discovering statistics using IBM SPSS Statistics*. London: Sage publication.

Jamali, D. (2004). Success and failure mechanisms of public private partnerships (PPPs) in developing countries: Insights from the Lebanese context. *International Journal Public Sector Management, 17*(5), 414–430.

Hu, Y., Chan, A. P., Le, Y., Xu, Y., & Shan, M. (2016). Developing a program organization performance index for delivering construction megaprojects in China: Fuzzy synthetic evaluation analysis. *Journal of Management in Engineering, 32*, 05016007.

Ismail, S. (2013a). Critical success factors of public private partnership (PPP) implementation in Malaysia. *Asia-Pacific Journal of Business Administration, 5*(1), 6–19.

Khosravi, S., & Afshari, H. (2011). *A success measurement model for construction projects*. International conference on financial management and economics. IPRD, Vol. 11. Singapore: IACSIT Press.

Lam, P., & Javed, A. (2015). Comparative study on the use of output specifications for Australian and U.K. PPP/PFI Projects. *Journal of Performance of Constructed Facilities*. https://doi.org/10.1061/(ASCE)CF.1943-5509.0000554.

Lam, E. W., Chan, A. P., & Chan, D. W. (2008). Determinants of successful design-build projects. *Journal of Construction Engineering and Management, 134*(5), 333–341.

Li, B., Akintoye, A., Edwards, P. J., & Hardcastle, C. (2005a). Critical success factors for PPP/PFI projects in the UK construction industry. *Construction Management and Economics, 23*(5), 459–471.

Li, B., Akintoye, A., Edwards, P. J., & Hardcastle, C. (2005b). Perceptions of positive and negative factors influencing the attractiveness of PPP/PFI procurement for construction projects in the UK: Findings from a questionnaire survey. *Engineering Construction Architectural Management, 12*(125), 148.

Li, B., Akintoye, A., Edwards, P. J., & Hardcastle, C. (2005c). The allocation of risk in PPP/PFI construction projects in the UK. *International Journal of project management, 23*(1), 25–35.

Lim, C. S., & Mohamed, M. Z. (1999). Criteria of project success: an exploratory re- examination. *International Journal of Project Management, 17*, 243–248.

Lingard, H., & Rowlinson, S. (2006). Letter to the editor. *Construction Management Economics, 24*(1107), 1109.

Liu, J., Love, P., Davis, P., Smith, J., & Regan, M. (2015). Conceptual framework for the performance measurement of public-private partnerships. *Journal of Infrastructure Systems*. https://doi.org/10.1061/(ASCE)IS.1943-555X.0000210.

Liyanage, C., & Villalba-Romero, F. (2015). Measuring success of ppp transport projects: a cross-case analysis of toll roads. *Transport Reviews, 35*(2), 140–161.

Meng, X., Zhao, Q., & Shen, Q. (2011). Critical success factors for transfer-operate-transfer urban water supply projects in China. *Journal of Management in Engineering*. https://doi.org/10.1061/(ASCE)ME.1943-5479.0000058.

Ministry of Finance and Economic Planning. (2011). *National policy on public – private partnerships*. Ghana.

Mladenovic, G., Vajdic, N., Wundsch, B., & Temeljotov-Salaj, A. (2013). Use of key performance indicators for PPP transport projects to meet stakeholders' performance objectives. *Built Environment Project and Asset Management, 3*(2), 228–249.

Norusis, M. J. (2008). *SPSS 15.0 guide to data analysis*. Upper Saddle River, NJ: Prentice Hall.

Osei-Kyei, R., & Chan, A. P. C. (2015a). Developing transport infrastructure in Sub-Saharan Africa through public–private partnerships: Policy practice and implications. *Transport Reviews*, 1–17.

Osei-Kyei, R., & Chan, A. P. (2017a). Factors attracting private sector investments in public–private partnerships in developing countries. *Journal of Financial Management of Property and Construction*.

Osei-Kyei, R., & Chan, A. P. (2017d). Developing a project success index for public–private partnership projects in developing countries. *Journal of Infrastructure Systems, 23*(4), 04017028.

Osei-Kyei, R., & Chan, A. P. (2018b). Comparative study of governments' reasons/motivations for adopting public-private partnership policy in developing and developed economies/countries. *International Journal of Strategic Property Management, 22*(5), 403–414.

Osei-Kyei, R., Dansoh, A., & Ofori–Kuragu, J. K. (2014). Reasons for adopting Public–Private Partnership (PPP) for construction projects in Ghana. *International Journal of Construction Management, 14*(4), 227–238.

Osei-Kyei, R., Chan, A. P. C., Javed, A. A., & Ameyaw, E. E. (2016). Critical success criteria for public-private partnership projects: International experts' opinion. *International Journal of Strategic Property Management, 21*.

Ozdoganm, I. D., & Birgonul, M. T. (2000). A decision support framework for project sponsors in the planning stage of build-operate-transfer (BOT) projects. *Construction Management and Economics, 18*(3), 343–353.

Parfitt, M. K., & Sanvido, V. E. (1993). Checklist of critical success factors for building projects. *Journal of Management in Engineering, 9*(3), 243–249.

Rosenau, M. D. (1984). *Project management for engineers, Lifetime Learning Pub*. Ann Arbor: University of Michigan.

Toor, R. S., & Ogunlana, S. O. (2010). Beyond the 'iron triangle': Stakeholder perception of key performance indicators (KPIs) for large-scale public sector development projects. *International Journal of Project Management, 28*(3), 228–236.

Westerveld, E. (2003). The Project Excellence Model: linking success criteria and critical success factors. *International Journal of Project Management, 21*(6), 411–418.

World Bank. (2009). *Attracting investors to African public private partnerships: A project preparation guide*. Washington, DC: World Bank Publications.

World Bank. (2014e). *Overcoming constraints to the financing of infrastructure: Success stories and lessons learned: Country, sector and project examples of overcoming constraints to the financing of infrastructure*. Washington, DC: World Bank Group.

Xu, Y., Yeung, J. F., Chan, A. P. C., Chan, D. W., Wang, S. Q., & Ke, Y. (2010b). Developing a risk assessment model for PPP projects in China—A fuzzy synthetic evaluation approach. *Automation in Construction, 19*(7), 929–943.

Yeung, J. F., Chan, A. P., & Chan, D. W. (2009). Developing a performance index for relationship-based construction projects in Australia: Delphi study. *Journal of Management in Engineering, 25*(2), 59–68.

Yuan, J., Zeng, A. Y., Skibniewski, M. J., & Li, Q. (2009). Selection of performance objectives and key performance Indicators in public-private partnerships projects to achieve value for money. *Construction Management and Economics, 27*(3), 253–270.

Zhang, X. (2006a). Public clients' best value perspectives of public private partnerships in infrastructure development. *Journal of Construction Engineering and Management*. https://doi.org/10.1061/(ASCE)0733-9364(2006)132%3A2(107).

Zhang, X. (2006b). Factor Analysis of Public Clients' Best-Value Objective in Public–Privately Partnered Infrastructure Projects. *Journal of Construction Engineering and Management*. https://doi.org/10.1061/(ASCE)0733-9364(2006)132:9(956).

Chapter 7
Management of Unsolicited Public-Private Partnership Projects

Abstract There are two major procurement routes for PPP projects; these are the traditional approach (solicited approach) and the unsolicited approach. Apparently, the traditional approach has been well documented and used across the world, however, the unsolicited has received little attention in the mainstream literature. In fact, many countries particularly those in the developing regions have increased their adoption of the unsolicited procurement method, meanwhile the unsolicited approach is tainted with allegations of corruption and nepotism. Further, it has been advocated that the unsolicited procurement method does not offer value for money to the public sector. Considering these, this chapter aims to holistically explore the salient issues in the management of unsolicited PPP projects. Specifically, the motivations behind its adoption and the effective ways of implementing it will be discussed using empirical evidence.

Keywords Unsolicited proposals · Private developers · Governments · Tendering methods · Unsolicited PPPs

7.1 Introduction

There are two major procurement routes for PPP projects; these are the traditional approach (solicited approach) and the unsolicited approach. Apparently, the traditional approach has been well documented and used across the world, however, the unsolicited has received little attention in the mainstream literature. In fact, many countries particularly those in the developing regions have increased their adoption of the unsolicited procurement method, meanwhile the unsolicited approach is tainted with allegations of corruption and nepotism. Further, it has been advocated that the unsolicited procurement method does not offer value for money to the public sector. Considering these, this chapter aims to holistically explore the salient issues in the management of unsolicited PPP projects. Specifically, the motivations behind its adoption and the effective ways of implementing it will be discussed using empirical evidence. The empirical evidence provided by Osei-Kyei et al.

R. Osei-Kyei, A. P. C. Chan, *International Best Practices of Public-Private Partnership*, https://doi.org/10.1007/978-981-33-6268-0_7

(2018a, b) will be used. Osei-Kyei et al. (2018a) conducted a questionnaire survey with experienced PPP practitioners across the world. The selected practitioners were requested to assess and rank the motivations for adopting unsolicited PPPs as well as the effective strategies that can be used to manage unsolicited PPPs.

7.2 Concept of Unsolicited Public-Private Partnership

In practice, PPP projects can be initiated through two main approaches: solicited and unsolicited (PPIAF 2014). The solicited approach is where the public agency/contracting authority identifies a project and invites private investors to submit their proposals. As a matter of fact, solicited PPP projects are procured through a well-established procurement system, which requires transparent and competitive tendering process (Neves and Kim 2017). The alternative is the unsolicited approach, which requires a private investor to identify and submit a project idea or concept to a public agency/department without any explicit invitation (Abdel Aziz and Nabavi 2014). Unlike solicited PPP projects, unsolicited PPPs can either be subjected to an open tendering process or directly negotiated with the original private proponent. But the latter is the method mostly adopted by many governments, and this makes unsolicited PPPs more complicated (Zawawi et al. 2016).

Typically, unsolicited PPP projects are not supposed to be identified in the public sector's infrastructure plan or list of projects, and they should have exclusive innovative ideas (Hodges and Dellacha 2007). In some situations, unsolicited proposals could be considered if there are no credible technical and financial solutions for a proposed project by the public sector. This is often seen in underground and tunnel projects (PPIAF 2009). Given the complicated issues of unsolicited proposals, particularly their non-competitive nature, many have failed to live up to expectations. Essentially, some countries, including the United Kingdom, do not consider the use of unsolicited proposals for PPP project implementation (PPIAF 2009). This is basically because of their controversial nature (Ballingall 2014). If not managed properly, unsolicited PPP projects could lead to countless agitations and demonstrations from the general public, poor value for money, not meeting the real societal and economic needs of the country, and satisfaction of few public and private officials (Abdel Aziz and Nabavi 2014; Ballingall 2014; Hodge 2003). Nonetheless, some countries, including South Korea, Chile, Argentina, Philippines, China, Taiwan, South Africa, Australia, and the United States (Virginia state) have adopted procedures and systems to manage the use of unsolicited PPP proposals (PPIAF 2009; Hodges and Dellacha 2007; Abdel Aziz and Nabavi 2014). Generally, the management procedures of unsolicited PPP proposals can be grouped into two categories (Hodges and Dellacha 2007): (1) procedures that relate to the approval and assessments of proposals, and (2) procedures that relate to the tendering process of proposals. Different countries have different requirements in the two procedures. For example, during the approval and assessment of unsolicited PPP proposals, countries such as Chile and Argentina require the private proponent to submit a general concept, after which, if approved, the private proponent would be asked to submit a detailed proposal within a stipulated period (Hodges and Dellacha 2007), whereas in South Africa and South Korea, a detailed proposal is only required for review (Yun et al. 2015; Hodges and Dellacha 2007). Also, countries such as Argentina require a bid bond to guarantee the seriousness of the private proponent (Hodges and Dellacha 2007).

Importantly, very few countries that have adopted unsolicited PPPs ensure that approved proposals undergo a competitive tendering process with some form of advantage or compensation to the original proponent (Hodge 2003; PPIAF 2009). Several competitive tendering methods exist, but the systems that are mostly adopted by governments include the bonus system, best and final offer system, Swiss challenge, and regular procurement (Hodges and Dellacha 2007; Zawawi et al. 2016; PPIAF 2014). The bonus system requires that a bonus point is given to the original proponent in an open tendering process (PPIAF 2014). This tendering system is commonly used in Chile and South Korea (Hodges and Dellacha 2007). Best and final offer is often used in South Africa and Argentina (Zawawi et al. 2016). This system involves a two-stage tendering process, in which the original proponent is automatically selected to compete with other bidders at the final stage. At the final stage, bidders, including the original proponent, are expected to submit their best and final offer for the proposed project (PPIAF 2014). Furthermore, in the best and final offer system, countries including South Africa and Chile ensure that the winning bidder compensates the original proponent for the development costs of the proposal, if the original proponent is not selected at the final round (Hodges and Dellacha 2007). In the Swiss challenge tendering system, the public agency/contracting authority invites other bidders to submit their proposals in an open tendering process. The original proponent wins the contract if willing to counter match any superior offer by other bidders (i.e., mostly the lowest-priced bidder) (Zawawi et al. 2016; PPIAF 2014). Apparently, if there is no superior offer from other bidders, the original proponent automatically wins the contract. This tendering system is commonly practiced in the Philippines, Italy, and Taiwan (Hodges and Dellacha 2007). In the Philippines, the original proponent is normally given 30 working days to match the price of the proposal from the leading bidder (Hodges and Dellacha 2007). For the regular procurement system, the contracting authority/public agency conducts an open tendering process, and the original proponent competes on equal terms with other bidders (PPIAF 2014). It should be mentioned that, in this system, the original proponent is compensated for the development rights and costs by the public authority before conducting an open tendering. Basically, each of the four commonly used tendering systems for managing unsolicited PPP proposals has its own pros and cons. However, if proper management strategies are adopted, irrespective of the tendering system used, the outcome will be highly positive and beneficial (Hodge 2003).

7.3 Data Collection

An international questionnaire survey was conducted with 400 identified PPP experts from the academic and/or industrial sectors. Questionnaires were sent to targeted respondents by emails with an option of responding on the "Survey Monkey" online questionnaire platform. After sending several reminders within four weeks, 57 valid responses were received. This represents an effective response rate of 14.25%. Apparently, this response rate is very low; however, the sample size of 57 is considered satisfactory for further analysis when compared to similar international survey studies in PPP; for example, Osei-Kyei et al. (2017) obtained 42 responses out of 310, Ameyaw and Chan (2015a, b, c) obtained 35 responses out of 326 and Zhang (2005a, b) obtained 46 responses out of 200. Furthermore, as a general rule of thumb, the central limit theorem holds for samples greater than 30, thus considering that the sample size of this study is larger, statistical analysis can still be performed to draw useful conclusions for future reference (Hwang et al. 2017). Aside these, the extensive practical and/or research experience of respondents (i.e. around 63% of respondents have more than 10 years of experience in PPP) and the diverse cultural distribution of respondents (i.e. experts from 25 different countries) demonstrate the reliability and richness of the survey responses for further analysis. Summary of respondents' backgrounds (i.e. PPP experience and countries) are shown in Tables 7.1 and 7.2.

Table 7.1 Respondents' profile

Demographics	Category	No. of respondents	Percentage
Sector	Academics	21	35.84
	Industrial practitioners	36	63.16
	Total	57	100
Years of industrial/research experience in PPP	5 years and below	8	14.00
	6–10 years	13	22.80
	11–15 years	16	28.10
	15–20 years	11	19.30
	21 years and above	9	15.80
	Total	57	100.0
Position in organizations			
Industrial practitioners	Director	13	36.11
	Manager	11	30.56
	Project analyst/ Consultant	12	33.33
	Total	36	100.00
Academics	Professor	7	33.33
	Associate Professor	4	19.05
	Lecturer	5	23.81
	Research Fellow	5	23.81
	Total	21	100.00

Osei-Kyei et al. (2018a) *(With permission from ASCE)*

Table 7.2 Countries of respondents

Countries	No. of respondents	Percentage
Australia	10	17.54
U.S.A.	6	10.53
U.K.	4	7.02
Malaysia	4	7.02
Philippines	3	5.26
Indonesia	3	5.26
India	3	5.26
France	3	5.26
Netherlands	2	3.51
Hong Kong	2	3.51
Germany	2	3.51
Canada	2	3.51
U.A.E	1	1.75
Spain	1	1.75
Serbia	1	1.75
Portugal	1	1.75
Nigeria	1	1.75
Kenya	1	1.75
Italy	1	1.75
Ghana	1	1.75
Colombia	1	1.75
China	1	1.75
Bulgaria	1	1.75
Brazil	1	1.75
Austria	1	1.75
Total	57	100

Source: Osei-Kyei et al. (2018a) *(With permission from ASCE)*

As presented in Table 7.1, 36% of the total respondents are from the academic sector, whereas 63% are industrial practitioners. Further, 86% of the total respondents have 6 or more years of experience in PPP research and/or practice. Also, majority of responses of industrial practitioners are from directors (i.e. 36%) and project analysts/consultants (i.e. 33%). In addition, majority of respondents from the academic sector are professors. With respect to respondents' origin, experts from 25 different countries participated in the study. Clearly, the backgrounds of respondents demonstrate the reliability and genuineness of the findings from this study.

7.4 Motivations for Governments' Adoption of Unsolicited Public-Private Partnerships

The increasing adoption of unsolicited PPPs clearly shows that many governments believe that implementing PPP projects using unsolicited proposals are of great benefit and can offer unique contribution to public infrastructure development. Table 7.3 presents the mean scores of the motivations for adopting unsolicited proposals for PPP. To determine the criticality of each motivation, the normalisation method was adopted. Of the seven motivations that emerge as critical, four are considered very critical because their mean values exceed 4.00 (values range between 4.00 and 4.46). They are "enhanced private sector innovation and creativity in PPPs (M4)"; "lack of public sector capacity (i.e. financial and technical) to identify, prioritise and procure projects (M1)"; "lack of private investors/developers interest in projects at remote areas (M9"; and "rapid implementation of PPP projects (M2)".

Enhanced private sector innovation and creativity in PPPs (M4) is the most critical motivation for using unsolicited proposals for PPP implementation. It has a mean value of 4.46 and a normalised value of 1.00. Importantly, the findings of Public-Private Infrastructure Advisory Facility (PPIAF) (2014) also emphasise the enhancement of private sector innovation as the major motivation for many

Table 7.3 Criticality ranking of governments' motivations for adopting unsolicited proposals for PPP project

Codes	Governments' motivations for implementing PPP using unsolicited proposals	Overall respondents (N = 57)				
		Mean	N	Mean	Normalization	Rank
M4	Enhance private sector innovation and creativity in PPPs	4.43	57	4.46	1.00	1
M1	Lack of public sector capacity (i.e. financial and technical) to identify, prioritize and procure projects	4.29	57	4.37	0.95	2
M9	Lack of private investors/developers' interest in projects at remote areas	4.52	57	4.26	0.89	3
M2	Rapid implementation of PPP projects	4.14	57	4.00	0.73	4
M10	Quick access to private sector finance for projects	3.95	57	3.98	0.72	5
M3	Avoid lengthy competitive bidding process	4.10	57	3.91	0.68	6
M7	Save costs involved in competitive bidding process	4.05	57	3.82	0.63	7
M11	Enable corruption and nepotism by avoiding competition	4.33	57	3.58	0.49	8
M8	Protect the intellectual property rights of investors	3.38	57	3.39	0.38	9
M5	Absence of proper and comprehensive policy framework for PPPs	2.76	57	3.00	0.15	10
M6	Fear of entrusting the proposed project to a less qualified firm through competitive bidding	3.71	57	2.74	0.00	11

Source: Osei-Kyei et al. (2018b) *(With permission from Emerald Group Publishing Ltd)*

governments' interest in unsolicited PPP proposals. These outcomes are quite understandable because the unsolicited approach requires private proponents to develop ideas that are very unique and extraordinary (Public-Private Infrastructure Advisory Facility (PPIAF), 2009). More importantly, private proponents are expected to present very innovative solutions to real societal problems. However, the impact and usefulness of the private sector's innovation and creativity in unsolicited PPPs can only be realised, if contracting authorities thoroughly assess how proposed innovative solutions conform to national or sector policy. This will adequately enhance the societal benefits of private sector's creativity and innovation in unsolicited PPPs.

The second very critical motivation is "lack of public sector capacity (i.e. financial and technical) to identify, prioritise and procure projects (M1)". It has a high mean value of 4.37 and a normalised value of 0.95. This finding is also in line with the outputs of World Bank (2017), where "M1" was identified as a critical reason for many governments' interest in unsolicited PPP projects. Generally, before the implementation of a PPP project, the contracting authority evaluates the proposed project's viability and need analysis through pre-feasibility and feasibility studies (Ng et al. 2012). Normally, public institutions that have adequate funds often contract out this obligation to private firms often called Transaction Advisors. However, those that do not have the financial and technical capacities to conduct the preliminary studies usually tend to rely on the efforts and initiatives of private developers (Public-Private Infrastructure Advisory Facility (PPIAF) 2009). This is often seen in many developing countries particularly those in sub-Saharan Africa and South Asia (Public-Private Infrastructure Advisory Facility (PPIAF) 2014; Hodges and Dellacha 2007). It should be noted that though "M1" may be a genuine reason for public sector departments to accept unsolicited proposals, governments should still make efforts to strengthen the capacities of their public institutions so that they can negotiate effectively with private proponents. Importantly, this will help avoid the misallocation of risks and poor value for money often tainted with unsolicited PPP projects.

Lack of private investors/developers' interest in projects at remote areas (M9) is third very critical motivation for many governments' interest in unsolicited PPP projects. This motivation also has a high mean value of 4.26 and a normalised value of 0.89. Importantly, this motivation was also identified by Hodges (2003) as one of the major reasons why many governments engage in unsolicited PPP practice. Hodges and Dellacha (2007) explained that usually governments easily accept proposals from private developers which target projects at smaller municipalities because they believe that competition for the market in such areas is limited. PPP projects in remote areas particularly smaller municipalities are characterised with a plethora of risks considering many obstacles (Osei-Kyei and Chan 2017). Hence, some private investors usually develop proposals targeting projects in these smaller municipalities because they believe that many other developers may not be interested to invest in such communities (Hodges and Dellacha 2007). Although, public departments/agencies may accept unsolicited proposals based on this reason (i.e. M9), they have to provide additional support to private proponents so that projects that are developed at these remote areas can be affordable to local commuters. If governments allow private proponents to develop PPP projects in remote areas without any additional support, they can charge exorbitant user tariffs to cover their huge investment costs and risks.

The fourth very critical motivation is "rapid implementation of PPP projects (M2)". It has high mean and normalised values of 4.00 and 0.73, respectively. Essentially, this is one of the most reported motivations by many institutional literatures including Public-Private Infrastructure Advisory Facility (PPIAF) (2009), Public-Private Infrastructure Advisory Facility (PPIAF) (2014). This is unsurprising because usually the unsolicited approach of implementing PPP does not follow any structured planning process as compared to the solicited ones (Hodges 2003). It mostly avoids the lengthy competitive tendering process and therefore allows projects to be negotiated within a shorter period (Public-Private Infrastructure Advisory Facility (PPIAF) 2014). Also, in case of an urgent need of infrastructure, the unsolicited approach is often adopted to fulfil this urgent requirement. However, it should be emphasised that though the unsolicited route can facilitate the rapid implementation of PPP projects, if it is not properly managed, projects can delay for several years (Hodges and Dellacha 2007). In this regard, governments should adopt proper strategies such as thorough evaluation of proposals and streamline of approval process to achieve their objective.

7.4.1 Comparison of Rankings Among Experts from Developing and Developed Countries on the Motivations for Adopting Unsolicited Proposals for PPP projects

The perception of respondents from developing and developed countries is tested hypothetically using the independent two-sample t-test. This investigation was important because different governments particularly between developing and developed countries have different approaches of implementing PPP projects; thus, there is the need to further explore possible differences in motivations among these two divergent economic groups. The test is conducted with the null hypothesis that no significant difference(s) exists in the perception of respondents from the developing and developed countries on the criticality of motivations for adopting unsolicited PPP projects. The test is conducted at a significance test value of 0.05 (95% confidence interval), therefore a significance value less than 0.05 rejects the null hypothesis. Table 7.4 shows the results of the independent t-test. As shown in the table (i.e. Sig. [two-tailed]), three motivations are statistically significant suggesting that respondents from developing and developed countries share different views on the criticality of the three motivations. These motivations are "fear of entrusting the proposed project to a less qualified firm through competitive bidding", "lack of private investors/developers' interest in projects at remote areas" and "enable

Table 7.4 Results of Independent Two-Sample t-test for governments' motivations for adopting unsolicited PPP projects

Code	Governments motivations for PPP using unsolicited proposals	t-test for Equality of Means				
		t	df	Sig. (2-tailed)	Lower	Upper
M1	Lack of public sector capacity (i.e. financial and technical) to identify, prioritize and procure projects	0.912	55	0.366	−0.157	0.419
		0.885	38.171	0.382	−0.169	0.431
M2	Rapid implementation of PPP projects	−1.058	55	0.295	−0.655	0.202
		−1.178	54.139	0.244	−0.611	0.159
M3	Avoid lengthy competitive bidding process	−1.246	55	0.218	−0.756	0.176
		−1.386	54.107	0.172	−0.709	0.129
M4	Enhance private sector innovation and creativity in PPPs	0.251	55	0.803	−0.305	0.393
		0.244	38.474	0.809	−0.319	0.406
M5	Absence of proper and comprehensive policy framework for PPPs	1.649	55	0.105	−0.081	0.835
		1.761	50.356	0.084	−0.053	0.807
M6	**Fear of entrusting the proposed project to a less qualified firm through competitive bidding**	−9.518	55	0.000*	−1.873	−1.222
		−9.733	44.89	0.000	−1.868	−1.227
M7	Save costs involved in competitive bidding process	−1.725	55	0.090	−0.764	0.057
		−1.730	42.398	0.091	−0.765	0.059
M8	Protect the intellectual property rights of investors	0.032	55	0.974	−0.481	0.497
		0.030	32.089	0.976	−0.535	0.550
M9	**Lack of private investors/ developers' interest in projects at remote areas**	−2.239	55	0.029*	−0.782	−0.043
		−2.467	53.45	0.017	−0.748	−0.077
M10	Quick access to private sector finance for projects	0.206	55	0.837	−0.415	0.511
		0.224	52.327	0.823	−0.379	0.474
M11	**Enable corruption and nepotism by avoiding competition**	−5.618	55	0.000*	−1.621	−0.768
		−5.753	45.083	0.000	−1.613	−0.776

Source: Osei-Kyei et al. (2018b) *(With permission from Emerald Group Publishing Ltd)*

corruption and nepotism by avoiding competition". Fear of entrusting the proposed project to a less qualified firm through competitive bidding (M6) has a significance value of 0.633 from the Levene's test for equality of variances results. This implies that equal variances are assumed; therefore, the resulting t-test significance is 0.00 with t statistic value of −9.518. Apparently, respondents from developing countries rated it higher than those from the developed countries. The possible reason for this difference is that most of the private firms in developing countries particularly in sub-Saharan Africa have little experience in PPP project delivery. In this regard, most public departments/agencies accept new initiative by private proponents with the belief that they as developers of the project idea have a higher potential to deliver successfully. In fact, because the chances of the original proponent to win the contract under open competitive bidding can be slim, the proposed project is allowed to go through as unsolicited so that the project is not eventually contracted to other local private firms who the public departments believe may not have adequate techniques to deliver [Public-Private Infrastructure Advisory Facility (PPIAF) 2009]. On the contrary, this is not often the same in the developed countries. Most developed countries have well established local private companies who have the capacity to offer alternative innovative solutions to unsolicited proposals if they go through a competitive bidding procedure. Therefore, many governments in the developed countries may not be motivated to adopt unsolicited PPPs based on this reason (i.e. M6). From the Levene's test for equality of variances results (Table 7.4), "lack of private investors/developers' interest in projects at remote areas" has a significance of 0.690, which signifies that equal variances are assumed. Therefore, the resulting t-test significance is 0.029, with a t statistic value of −2.239. Usually, remote municipalities and districts in developing countries are very underdeveloped and lack basic public utilities compared to the developed countries. Hence, local authorities are keen to consider unsolicited proposals as a means of attracting private investors into PPP projects in these remote areas (Hodges 2003).

"Enable corruption and nepotism by avoiding competition" has a Levene's test significance value of 0.965, implying that equal variances are assumed. Therefore, the resulting t-test significance value is 0.00, with a t statistic value of −5.618. This outcome supports the findings of World Bank (2017) where it was found out that, public officials particularly those in developing countries allow the use of unsolicited PPPs to engage in fraudulent practices. Precisely, in many developing countries particularly those in sub-Saharan Africa and South Asia, unsolicited PPPs has become an enabling corruption device for corruption due to their lack of competition and inadequate transparency, and this has been reported in many past projects (World Bank 2017; Osei-Kyei and Chan 2017). It is therefore not surprising that the respondents from developing countries ranked it very high. Although, there could be allegations of corruption with the use of unsolicited PPPs in some developed countries, certainly, it is not as extensive as seen in the developing countries. Notwithstanding, many developed countries have established proper measures and systems which deal with allegations of corruptions compared to many developing countries.

7.5 Effective Strategies for Managing Unsolicited PPP Proposals

It is necessary for practitioners and policymakers to be fully aware of the strategies to adopt to successfully manage unsolicited proposals for PPP project implementation. The ranking and mean significance index of strategies for managing unsolicited proposals are presented in Table 7.5. For a strategy to be considered critical, it should have a mean value/index equal to or above 3.50 (Ahadzie et al. 2008). Importantly, using this criticality cut off point (3.50) in this study ensured that strategies selected for further discussion are not relatively neutral (i.e., neither important nor unimportant) but rather significant and essential to practice (Ahadzie et al. 2008). As presented in Table 7.5, seven strategies emerge as critical because their mean values range between 3.58 and 4.39. Furthermore, standard deviations of the seven strategies range from 0.56 to 0.90. These strategies include existence of well-structured and clear policy guidelines for unsolicited proposals (S6); thorough assessment of value for money, innovation, cost, and risks of proposals (S1); employment of highly skilled and competent staff during evaluations of proposals (S4); competitive, fair, and transparent tendering process (S8); extensive public

Table 7.5 Criticality ranking of strategies for effective management of unsolicited PPP proposals

Code	Strategies for effective management of unsolicited PPP proposals	Overall Respondents			
		N	Mean	Normalization values	Rank
S6	Existence of well-structured and clear policy guidelines for unsolicited proposals	57	4.39	1.00	1
S1	Thorough assessment of value for money, innovation, cost and risks of proposals	57	4.37	0.99	2
S4	Employment of highly skilled and competent staff during evaluations of proposals	57	4.19	0.90	3
S8	Competitive, fair and transparent tendering process	57	4.05	0.83	4
S5	Extensive public consultation and stakeholder engagement	57	3.74	0.67	5
S2	Comprehensive evaluation of the impact of unsolicited proposals on sector/national policy	57	3.61	0.60	6
S12	Adequate protection of intellectual property rights of the original proponent	57	3.58	0.58	7
S11	Effective coordination among government agencies and authorities	57	3.44	0.51	8
S3	Streamline of approval process for proposals	57	3.30	0.44	9
S7	Expedited review and evaluation process	57	3.18	0.38	10
S9	Reasonable amount of reimbursement for project development costs to original proponent	57	2.51	0.04	11
S10	Adequate government incentives and guarantees for unsolicited PPP projects	57	2.44	0.00	12

Source: Osei-Kyei et al. (2018a) *(With permission from ASCE)*

consultation and stakeholder engagement (S5); comprehensive evaluation of the impact of unsolicited proposals on sector/national policy (S2); and adequate protection of intellectual property rights of the original proponent (S12). This outcome suggests that the seven critical strategies need continuous and careful attention by practitioners and policymakers to manage unsolicited PPP proposals and achieve positive outcomes.

The most critical strategy is existence of well-structured and clear policy guidelines for unsolicited proposals (S6). It has a very high mean value of 4.39. This finding is unsurprising because a clear regulatory framework and policy guide have been reported by many researchers (including Cheung et al. (2012a, b), Osei-Kyei and Chan (2015b), Li et al. (2005), and Abdel Aziz (2007)) as very essential to the successful management of PPP projects in general. Apparently, this outcome also reinforces assertion by the PPIAF (2009) and Hodges and Dellacha (2007) that without the existence of a clear policy guide and framework for unsolicited PPP proposals, many unsolicited PPP projects are bound to face many obstacles. As a matter of fact, a well-structured unsolicited proposal policy ensures that private investors submit reasonable and innovative proposals; more importantly, it enables public departments to receive proposals that are in line with national interest (PPIAF 2014; Hodges and Dellacha 2007). A well-structured policy guide for unsolicited proposals should be explicit on the various public institutions responsible for handling the approvals of unsolicited proposals (PPIAF 2014). Furthermore, it has to create more transparency in the management of unsolicited PPP proposals. It is also important to note that establishing a clear policy guide for unsolicited proposals may not be enough to ensure the effective management of unsolicited proposals; however, the government has to ensure that the policy guide is consistently enforced by public departments (PPIAF 2014). The second critical strategy is thorough assessment of value for money, innovation, cost, and risks of proposals (S1). It has a mean score of 4.37 and a standard deviation value of 0.56. Generally, many governments tend to adopt unsolicited proposals for PPP implementation because they believe unsolicited proposals offer unique private sector innovations and value for money (Hodge 2003). However, it should be emphasized that innovation and value for money can only be sufficiently achieved if unsolicited proposals are thoroughly assessed (Yun et al. 2015; PPIAF 2009). Risks and costs are two key parameters that possibly affect the innovation and value for money aspects of unsolicited PPP projects; therefore, they have to be properly examined as well. As explained by PPIAF (2009), during proposal assessment, public departments/agencies should ensure that the value for money of the proposed PPP project is compared to other service delivery options. Furthermore, the viability of technology proposed by the proponent, conformity of risk allocation mechanism with national policy, and the estimated costs and financing plan should be ensured to be reasonable and sound (PPIAF 2009). Employment of highly skilled and competent staff during evaluations of proposals (S4) is the third critical strategy with a mean value of 4.19. In essence, the lack of technical and financial capacities of many public departments/agencies often makes it difficult for them to properly ascertain the viability of proposals submitted by private investors (World Bank 2017). This often leads to the distress or even failure of most unsolicited PPP projects (Hodges and Dellacha 2007). It is therefore very essential that public departments/agencies engage highly experienced professionals as part of their evaluation team when assessing unsolicited proposals (PPIAF 2009). Most proposals submitted by private investors are very technically advanced. Specifically, unsolicited proposals usually have unique techniques and applications that need very experienced experts to ascertain whether the proposed techniques are feasible and cost-efficient (PPIAF 2009).

The fourth critical strategy for the effective management of unsolicited PPP proposals is competitive, fair, and transparent tendering process (S8). It has a mean value of 4.05 and a standard deviation value of 0.83. Many past research studies have also pointed out the criticality of a competitive and transparent tendering process in the successful management of PPP projects in general (Cheung et al. 2012a, b; Zhang 2005a, b). This therefore reaffirms the fact that, irrespective of the approach for implementing PPP projects, competition and transparency are very necessary at the tendering stage (Osei-Kyei and Chan 2015b; Zawawi et al. 2016). Very often, unsolicited proposals are sole-sourced (Hodges and Dellacha 2007), and this has therefore contributed largely to the negative public perception about its usage (Hodge 2003). Emphatically, the introduction of competition and transparency in the management of unsolicited PPP proposals will enable the public sector to select the most cost-effective bid and, more importantly, agree on contractual terms that are in the general public's interest (PPIAF 2014). In recent years, many countries, including South Korea, Argentina, South Africa, Chile, and the Philippines, have introduced competitive and transparent tendering procedures into the management of unsolicited PPP proposals (Zawawi et al. 2016), and this has greatly improved the effectiveness in the use of unsolicited proposals for PPP project implementation (Yun et al. 2015). However, it should also be emphasized that the level of competition and transparency in the management of unsolicited proposals largely depends on the host government's policy (Hodge 2003). Nevertheless, it is crucial for governments to ensure that whatever level of competition and transparency they decide to use, it has to be clearly stated in the policy guide or framework (Hodges and Dellacha 2007).

Extensive public consultation and stakeholder engagement (S5) is the fifth critical strategy for the effective management of unsolicited PPP proposals. It has mean and standard deviation values of 3.74 and 0.88, respectively. Generally, public consultation and stakeholder engagement cannot be underestimated when implementing PPP projects. In this regard, a large strand of literature has identified this strategy to be very critical in the successful implementation of PPP projects in general (Zhang 2005a, b; Nisar 2013; Cheung et al. 2012a, b). However, it should be noted that this strategy (S5) needs more attention when unsolicited proposals are used for PPP implementation (PPIAF 2009). Many public agitations and political unrests over PPP projects, particularly the unsolicited ones, usually stem from the absence of extensive public and external stakeholder engagement (Osei-Kyei and Chan 2015a). Often, the general public and opposing political parties suspect some fraudulent acts with the awarding of contracts, although the government may have conducted a competitive tendering process. Apparently, it is the failure of the government to extensively engage the general public and interested parties in the proposed projects that mostly leads to these public agitations and political unrests. Therefore, at the early stages of the PPP process, the contracting authority should identify and incorporate all interested parties to avoid future opposition (PPIAF 2014).

The sixth critical strategy is comprehensive evaluation of the impact of unsolicited proposals on sector/national policy (S2). It also has mean and standard deviation values of 3.61 and 0.90, respectively. As emphasized by the PPIAF (2009), many public departments/agencies normally do not properly assess whether proposals submitted by private individuals have positive or negative impacts on the sector/national policy in terms of its infrastructure development. This deficiency often results in many unsolicited PPPs having a trivial positive impact on the infrastructure goals of the sector ministry and host country. In essence, public departments at the preliminary stages have to evaluate whether proposals submitted by private sector proponents address national/sector policy priorities (PPIAF 2009). Indeed, an unsolicited proposal should not be considered on the basis that there exists no such public facility, but proposals that meet the developmental goals of the sector ministry and the host country should be given consideration (PPIAF 2009).

Adequate protection of intellectual property rights of the original proponent (S12) is the last critical strategy for effective management of unsolicited PPP proposals. Unarguably, the proprietary rights of an unsolicited proposal belong to the private proponent, and it is the responsibility of the public authority to adequately protect the intellectual rights of the proponent if the unsolicited proposal is accepted (PPIAF 2014). Also, if a public department/agency decides to adopt a competitive and transparent tendering process, it has to ensure that it does not violate the proprietary rights of the investor by exposing the investor's techniques (Hodges and Dellacha 2007). Ideally, a licensing arrangement can be set up to acquire the proprietary techniques of the proposed project, if the original proponent techniques are needed. Also, the public department/agency can include the development costs of the project in the tender documents to compensate the original proponent if not selected as the final bidder (Hodges and Dellacha 2007). Importantly, these measures can help contracting authorities to avoid violating the proprietary rights of private proponents, particularly when unsolicited proposals go through a competitive tendering process. Protecting the proprietary rights of private proponents encourages more investors to submit unique ideas and techniques, which will be beneficial to the development of PPPs through the unsolicited approach.

7.6 Commonly Used and the Most Suitable Tendering Methods for Managing Unsolicited PPP Proposals

Table 7.6 presents results of the most commonly used and most suitable tendering methods in managing unsolicited PPP proposals. From Table 7.6, 50.9% of the total respondents indicated that regular procurement is what they have experienced or

Table 7.6 Commonly used tendering method for managing unsolicited PPP proposals

Tendering systems/ methods	Developed countries		Developing countries		Overall respondents	
	Frequency	Percentage	Frequency	Percentage	Frequency	Percentage
Swiss Challenge	3	8.3	1	4.8	4	7
Bonus System	7	19.4	1	4.8	8	14
Best and Final offer	5	13.9	1	4.8	6	10.5
Regular procurement	14	38.9	15	71.4	29	50.9
None	7	19.4	3	14.3	10	17.5
Total	36	100	21	100	57	100

Source: Osei-Kyei et al. (2018a) *(With permission from ASCE)*

used before in managing unsolicited PPP proposals. Also, 14% of the total respondents mentioned bonus system, whereas 7 and 10.5% mentioned Swiss challenge and best and final offer, respectively.

Essentially, 17.5% of the total respondents indicated that they have not used or experienced any of the given tendering methods. In contrast, majority of the total respondents (i.e., 47%) indicated that best and final offer should be the most suitable tendering method, whereas regular procurement should not be encouraged to be used for managing unsolicited PPP proposals. Also, Swiss challenge is indicated as the second most suitable tendering method for managing unsolicited PPP proposals. These outcomes support assertions by PPIAF (2009) and Hodges and Dellacha (2007) that the Swiss challenge and best and final offer are encouraged to be used by governments because they are relatively easy to define, and more importantly, they do not create opportunity for subjective decision making compared to the bonus system.

One possible reason why regular procurement is often discouraged is because mostly, the contracting authority pays a fee for the proprietary rights and initial development costs of the proposed project before inviting bidders to compete with the original proponent on equal terms (PPIAF 2014). Apparently, acquiring the development rights and ascertaining the exact initial development costs are very challenging, and more essentially, they create additional work for the contracting authorities (PPIAF 2009). These notwithstanding, best and final offer and Swiss challenge are the most adopted tendering methods by many countries that have set up well-structured procedures for managing unsolicited PPP proposals (Hodges and Dellacha 2007). It should be mentioned that these tendering methods are some of the measures adopted to ensure competition and transparency in the management of unsolicited proposals; therefore, they are very necessary for the effective management and delivery of unsolicited PPPs.

7.7 Chapter Summary

This chapter has explored the main issues in the management of unsolicited PPP proposals, which are the motivations for its adoption and the management strategies available. The discussion was based on empirical evidence provided from an international questionnaire survey with experienced experts. It was ascertained that the key reasons for the adoption of unsolicited PPP proposals for PPP Implementation include "enhance private sector innovation and creativity in PPPs (M4)", "lack of public sector capacity (i.e. financial and technical) to identify, prioritize and procure projects (M1)", "lack of private investors/developers interest in projects at remote areas (M9" and "rapid implementation of PPP projects (M2)". Further, the critical effective management strategies for unsolicited PPP proposals include existence of well-structured and clear policy guidelines for unsolicited proposals (S6), thorough assessment of value for money, innovation, cost and risks of proposals (S1), employment of highly skilled and competent staff during evaluations of proposals (S4), competitive, fair and transparent tendering process (S8), extensive public consultation and stakeholder engagement (S5), comprehensive evaluation of the impact of unsolicited proposals on sector/national policy (S2), adequate protection of intellectual property rights of the original proponent (S12) and effective coordination among government agencies and authorities (S11). Also, "regular procurement" is found to have been the commonly used tendering method for managing unsolicited proposals by the respondents. However, majority of the total respondents indicate that "best and final offer" and "Swiss challenge" are the best tendering methods for managing unsolicited PPP proposals. These outputs will inform international private investors and governments who are yet to adopt the unsolicited PPP route of the measures and systems that need to be considered in order to ensure the successful implementation of unsolicited PPP projects.

References

Abdel Aziz, A. M. (2007). Successful delivery of public-private partnerships for infrastructure development. *Journal of Construction Engineering and Management, 133*(12), 918–931.

Abdel Aziz, A., & Nabavi, H. (2014). Unsolicited proposals for PPP projects: Private sector perceptions in the USA. In *Construction Research Congress 2014: Construction in a global network* (pp. 1349–1358). Reston: American Society of Civil Engineers.

Ahadzie, D. K., Proverbs, D. G., & Olomolaiye, P. O. (2008). Critical success criteria for mass house building projects in developing countries. *International Journal of Project Management, 26*, 675–687.

Ameyaw, E. E., & Chan, A. P. (2015a). Risk allocation in public-private partnership water supply projects in Ghana. *Construction Management and Economics, 33*(3), 187–208.

Ameyaw, E. E., & Chan, A. P. (2015b). Evaluation and ranking of risk factors in public–private partnership water supply projects in developing countries using fuzzy synthetic evaluation approach. *Expert Systems with Applications, 42*(12), 5102–5116.

Ameyaw, E. E., & Chan, A. P. C. (2015c). Risk ranking and analysis in PPP water supply infrastructure projects: an international survey of industry experts. *Facilities, 33*(7/8), 428–453.

Ballingall, J. (2014). *PPPs: Fiscal space and investment: The United Kingdom*. Infrastructure U.K. http://idbdocs.iadb.org/wsdocs/getdocument.aspx?docnum=38831864i. Mar 2017.

Cheung, E., Chan, A. P. C., & Kajewski, S. (2012a). Factors contributing to successful public private partnership projects, Comparing Hong Kong with Australia and the United Kingdom. *Journal of Facilities Management, 10*(1), 45–58.

Cheung, E., Chan, A. P. C., Lam, P. T. I., Chan, D. W. M., & Ke, Y. (2012b). A comparative study of critical success factors for public private partnerships (PPP) between Mainland China and the Hong Kong Special Administrative Region. *Facilities, 30*(13/14), 647–666.

Hodge, J. (2003). *Unsolicited proposals: The issues of private infrastructure projects*. Washington, DC: World Bank Publications. http://siteresources.worldbank.org/EXTFINANCIALSECTOR/Resources/282884-1303327122200/257Hodge-031103.pdf. Mar 2017.

Hodges, J., & Dellacha, G. (2007). *Unsolicited infrastructure proposals: How some countries introduce competition and transparency PPIAF working paper no. 1, 2007*. http://www-wds.worldbank.org/external/default/WDSContentServer/WDSP/IB/2007/12/06/000020439_20071206143329/Rendered/PDF/417300Unsolici1alsOPPIAF0101 PUBLIC1.pdf. Mar. 2017.

Hwang, B. G., Zhu, L., & Tan, J. S. H. (2017). Identifying critical success factors for green business parks: case study of Singapore. *Journal of Management in Engineering, 33*(5). http://ascelibrary.org/doi/full/10.1061/%28ASCE%29ME.1943-5479.0000536.

Li, B., Akintoye, A., Edwards, P. J., & Hardcastle, C. (2005). Critical success factors for PPP/PFI projects in the UK construction industry. *Construction Management and Economics, 23*(5), 459–471.

Neves, P., & Kim, D. J. (2017). *Managing unsolicited proposals in infrastructure: 5 key questions for governments*. World Bank Publication. http://blogs.worldbank.org/ppps/managing-unsolicited-proposals-infrastructure-5-key-questions-governments?CID=PPP_E_PPPNewsletter_EXTi

Ng, S. T., Wong, Y. M., & Wong, J. M. (2012). Factors influencing the success of PPP at feasibility stage – A tripartite comparison study in Hong Kong. *Habitat International, 36*(4), 423–432.

Nisar, T. M. (2013). Implementation constraints in social enterprise and community Public Private Partnerships. *International Journal of Project Management, 31*(4), 638–651.

Osei-Kyei, R., & Chan, A. P. C. (2015a). Developing transport infrastructure in Sub-Saharan Africa through public–private partnerships: Policy practice and implications. *Transport Reviews*, 1–17.

Osei-Kyei, R., & Chan, A. P. C. (2015b). Review of studies on the critical success factors for public–private partnership (PPP) projects from 1990 to 2013. *International Journal of Project Management, 33*(6), 1335–1346.

Osei-Kyei, R., & Chan, A. P. (2017). Risk assessment in public-private partnership infrastructure projects: empirical comparison between Ghana and Hong Kong. *Construction Innovation*, 204–223.

Osei–Kyei, R., Chan, A. P. C., Javed, A. A., & Ameyaw, E. E. (2017). Critical success criteria for public-private partnership projects: International experts' opinion. *International Journal of Strategic Property Management, 21*. https://doi.org/10.3846/1648715X.2016.1246388.

Osei-Kyei, R., Chan, A. P., Dansoh, A., Ofori-Kuragu, J. K., & Oppong, G. D. (2018a). Strategies for effective management of unsolicited public–private partnership proposals. *Journal of Management in Engineering, 34*(3), 04018006.

Osei-Kyei, R., Chan, A. P., Dansoh, A., Ofori-Kuragu, J. K., & Owusu, E. K. (2018b). Motivations for adopting unsolicited proposals for public-private partnership project implementation: a survey of international experts. *Journal of Financial Management of Property and Construction*, 221–238.

PPIAF (Public-Private Infrastructure Advisory Facility). (2009). *"Unsolicited proposals." Module 5: Implementing and Monitoring*. Washington, DC: PPIAF. https://ppiaf.org/sites/ppiaf.org/files/documents/toolkits/highwaystoolkit/6/pdf-version/5-42.pdf.

PPIAF (Public-Private Infrastructure Advisory Facility). (2014). Unsolicited proposals – An exception to public initiation of infrastructure PPPs. Washington, DC: PPIAF. http://ppp.

worldbank.org/public-private-partnership/sites/ppp.worldbank.org/files/documents/PPIAF_UnsolicitedProposals_EN.pdf. Mar 2017.

Public-Private Infrastructure Advisory Facility (PPIAF). (2009). *Toolkit for Public-Private Partnerships in roads & highways: N4 Toll Road from South Africa to Mozambique*. Available from: http://www.ppiaf.org/sites/ppiaf.org/files/documents/toolkits/highwaystoolkit/6/pdf-version/safricamozambique.pdf

World Bank. (2017). *Guidelines for the development of a policy for man-aging unsolicited proposals in infrastructure projects*. Washington, DC. http://pubdocs.worldbank.org/en/976121488983070966/Guidelines-for-the-Development-of-a-USP-Policy.pdf. May 2017.

Yun, S., Jung, W., Han, S. H., & Park, H. (2015). Critical organizational success factors for public private partnership projects – A comparison of solicited and unsolicited proposals. *Journal of Civil Engineering and Management, 21*(2), 131–143.

Zawawi, M. I. Z., Kulatunga, U., & Thayaparan, M. (2016). Malaysian experience with public-private partnership (PPP): Managing unsolicited proposal. *Built Environment Project and Asset Management, 6*(5), 508–520.

Zhang, X. Q. (2005a). Criteria for selecting the private-sector partner in public–private partnerships. *Journal of Construction Engineering and Management, 131*(6), 631–644.

Zhang, X. Q. (2005b). Critical success factors for public-private partnerships in infrastructure development. *Journal of Construction Engineering and Management, 131*(1), 3–14.

Chapter 8
Conflict Management in Public-Private Partnership Projects

Abstract Conflict is certainly inevitable in public-private partnership (PPP) due to the long-term agreement and multiplicity of stakeholders with varying beliefs and interests. However, the thorough understanding of how conflict could be managed among project parties and stakeholders can help minimize the occurrence of it in PPP project arrangement. This will thereby reduce the negative repercussions of conflict occurrence in PPP arrangements. Considering these, this chapter discusses comprehensively on the management of conflict in PPP by focusing on its root causes, resolution mechanisms and prevention strategies.

Keywords Conflict management · Conflict prevention · Stakeholders · Conflict resolution methods · China · Ghana

8.1 Introduction

Conflict is certainly inevitable in public-private partnership (PPP) due to the long-term agreement and multiplicity of stakeholders with varying beliefs and interests. However, the thorough understanding of how conflict could be managed among project parties and stakeholders can help minimize the occurrence of it in PPP project arrangement. This will thereby reduce the negative repercussions of conflict occurrence in PPP arrangements. Considering these, this chapter discusses comprehensively on the management of conflict in PPP by focusing on its root causes, resolution mechanisms and prevention strategies. Discussion of these issues are done based on empirical evidence provided by Osei-Kyei et al. (2019a, b). In both studies, an empirical questionnaire survey was conducted with PPP practitioners from Ghana and China. Targeted practitioners were asked to rate on a five-point Likert scale the root causes of conflict and conflict prevention measures/strategies for PPP projects.

R. Osei-Kyei, A. P. C. Chan, *International Best Practices of Public-Private Partnership*, https://doi.org/10.1007/978-981-33-6268-0_8

8.2 Concept of Conflict Management in Public-Private Partnerships

In the construction industry, conflict is inevitable and remains a key risk because of the multiple players with different interests and values (Mitkus and Mitkus 2014; Ejohwomu et al. 2016). As defined by Gardiner and Simmons (1992), conflict is any divergence of interests, objectives or priorities between individuals, groups or organization; or non-conformance to requirement of a task, action or process. Essentially, conflict arises when there are serious differences in two or more beliefs, interests and perceptions (Acharya et al. 2006). Gorse (2003) put conflict into two categories; functional and dysfunctional. Functional conflict is natural, and it occurs when there are challenges and disagreements on tasks, roles, process and functions. On the contrary, dysfunctional conflict is considered unnatural and results from personal insults and criticisms that boost self-ego instead of improving task performance (Gorse 2003; Gardiner and Simmons 1992).

Although, all construction projects are prone to conflict, Public-Private Partnership (PPP) projects are more susceptible to conflict compared to any other construction project arrangement including the traditional-bid build system (Osei-Kyei and Chan 2015). This is because PPP projects involve a long-term arrangement with myriad of complex legal arrangements and a large number of stakeholders with multifaceted interests, beliefs and objectives (UNDP 2017). In addition, the multiplicity of external stakeholders of PPP projects exposes them highly to conflict problems. Projects that have suffered some form of distress due to conflicts include the Ghana National Housing Project (Ghana); Bangkok Elevated Transport System (BETS) (Thailand); West Cultural Kowloon District (Hong Kong) and The Ngone bridge project (Lao PDR). Indeed, conflict can be damaging to a project if not properly handled. It could lead to project delays, undetermined team spirit, increase in project cost and breaks personal and professional relationships. Because of the detrimental effects of conflict, it should therefore be properly managed by project parties (Harmon 2003).

The first step toward effective conflict management is to identify the root sources of conflicts (Jaffar et al. 2011; Osei-Kyei et al. 2018a, b). This enables the project manager to develop an effective strategic plan on how to deal and address any conflict in the project lifecycle. Many past studies have identified several potential sources of conflicts and these include: poor communication, inadequate design, limited resources, unrealistic client expectations, omissions in the contract documents and deficient management (Kumaraswamy 1997; Harmon 2003). In conflict management, adopting an effective resolution method is also essential (Osei-Kyei et al. 2018a, b). If conflict arises and it is not resolved with proper resolution mechanism within a specific timeframe, it grows larger into a full-scale dispute (Harmon 2003). Some commonly used conflict resolution mechanism in construction includes mediation, arbitration and negotiations (Chan and Suen 2005). However, it should be mentioned that conflict should be monitored and evaluated after a resolution mechanism is implemented or applied. This will help to assess the effectiveness of the resolution mechanism adopted and prevent similar conflict occurring (Harmon 2003).

Usually, conflict management is directly related to risks, as most of the sources of conflict could be risk factors. In this regard, conflict management must encompass risk management; risks should be properly identified, analysed and fairly allocated.

8.3 Data Collection

An empirical questionnaire survey was conducted with experienced practitioners in both China and Ghana. Questionnaires were sent to the targeted respondents by email and/or face-to-face meeting. Questionnaires distributed in China specifically in Chengdu and Yibin (Sichuan Province) were through face-to-face meetings, whereas questionnaires distributed to respondents in other cities (i.e. Beijing and Qingdao) were by emails. For Ghana, questionnaires were sent to all prospective respondents through emails. In total, 84 valid replies were received; 52 from Ghana and 34 from China. These represent response rates of 34.67% and 18.82% for Ghana and China respectively. Although the response rates were considerably low, the samples are considered satisfactory for further analysis considering the rich background of respondents. The detail background of respondents from both countries is shown in Table 8.1.

From Table 8.1, approximately 78% and 79% of respondents from China and Ghana are industrial practitioners; thus, the majority of respondents from both jurisdictions are exposed to the actual intricacies of PPP projects, which enriches the responses and enhances the reliability of the research outcomes. Furthermore, almost 66% of the respondents from China have more than 10 years of industrial

Table 8.1 Profile of respondents from China and Ghana

	Ghana		China	
Demographics	No. of replies	Percent (%)	No. of replies	Percent (%)
Sector of PPP				
Research/academic	11	21.2	7	21.9
Public sector organizations	24	46.2	13	40.6
Private institutions	17	32.7	12	37.5
Total	52	100	32	100
Years of experience				
≤10	27	51.9	11	34.4
11–20 years	24	46.2	20	62.5
≥21 years	1	1.9	1	3.1
Total	52	100	32	100

Source: Osei-Kyei et al. (2019a) (*With permission from Elsevier*)

and/or research experience in PPPs compared to from Ghana. This implies that more respondents from China have more PPP experience than their Ghanaian counterparts. This is quite unsurprising considering the fact that since 2000s there has been a tremendous increase in PPP investments in China (World Bank 2015a, b, c). Specifically, as at 2016, PPP investments in China has increased to a total of US$ 11.5 billion (World Bank 2017). Therefore, it is likely that more practitioners have become experienced in PPP. Nonetheless, because PPP is still developing in Ghana, the fairly high number of respondents with more than 10 years of experience suggests that opinions received from Ghana are credible.

8.4 Root Causes of Conflict in Public-Private Partnerships

Table 8.2 presents the rankings of the causes of conflicts in PPPs. As shown in the table, the mean values range from 2.83 to 4.62 and 2.66 to 4.44 for Ghana and China respectively. These outcomes suggest that the Ghanaian respondents rated the set of causes of conflicts slightly higher than their Chinese counterparts. To determine the critical causes of conflicts within each jurisdiction, the one-sample t-test was conducted. The test was conducted with the hypothesised value (U_0) set at 3.0. Of the 16 causes of conflict, 13 and nine emerged critical in Ghana and China respectively. The large number of causes of conflicts in PPPs emerging as critical in Ghana was anticipated because since the PPP concept became operationalized in 2011, several projects have been distressed and failed due to many misunderstandings and disputes between stakeholders. On the other hand, China has practiced PPP for quite a long time; therefore, they have rationalized their PPP process, which is very crucial in minimizing conflicts. Hence this may have contributed to the low number of causes emerging as critical.

The test results on the significant differences between Ghana and China is shown in the last column in Table 8.2. The test was conducted using the Mann-Whitney U statistic at a significance level of 0.05. According to the table, eight causes of conflicts in PPPs have significance test value below 0.05. This suggests that respondents from Ghana and China have different views and opinions on the rankings of the eight causes of conflicts. Essentially, this research outcome reinforces assertions that different countries have different features and conditions in their PPP markets, thus it is essential for PPP practices to be explored and compared among different countries/cultures in order to enrich the international best practice framework (Cheung et al. 2012a, b).

Four of the eight causes of conflicts with significant differences are ranked higher in Ghana and lower in China. They are political interference, delay in decision makings by parties, unrealistic time targets and inadequate compensation to displaced persons. These causes directly relate to poor governance and contractual arrangements. Political interference is ranked second by the Ghanaian respondents whereas, their Chinese counterparts ranked it 12th. Political interference in PPPs is very common in many underdeveloped countries such as Ghana, therefore is not surprising

Table 8.2 Mean analysis and significant test results for the causes of conflicts in PPPs in Ghana and China

		Ghana			China			Mann-Whitney U test		
Codes	Causes of conflicts in PPP	Mean	Rank	One sample T-test (Sig.)	Mean	Rank	One sample T-test (Sig.)	U statistics	Z	Sig.
C1	Unfair risk allocation	3.46	11	0.00	4.44	1	0.00	338	−4.865	0.00*
C2	Absence of proper communication channel	3.83	5	0.00	4.31	3	0.00	549.5	−2.898	0.00*
C3	Double meanings in output specifications	3.73	6	0.00	3.84	6	0.00	774	−0.572	0.57
C4	Lack of understanding on the roles and responsibilities of parties	4.38	3	0.00	4.41	2	0.00	800.5	−0.324	0.75
C5	Unexpected tariff changes	3.27	13	0.03	3.94	5	0.00	504.5	−3.179	0.00*
C6	Excessive contract variations	4.00	4	0.00	3.78	7	0.00	736	−1.020	0.31
C7	Political interference	4.40	2	0.00	3.09	12	0.54	224	−5.871	0.00*
C8	Ambiguous goals and objectives	3.54	10	0.00	3.56	8	0.00	785	−0.475	0.64
C9	Incomplete transfer of risks	3.19	14	0.11	4.25	4	0.00	339.5	−4.757	0.00*
C10	Personality clashes	2.83	16	0.18	2.97	13	0.88	749	−0.798	0.43
C11	Unrealistic time targets	3.60	8	0.00	2.94	15	0.69	496.5	−3.239	0.00*
C12	Delay in decision makings by parties	4.62	1	0.00	3.13	11	0.44	170.5	−6.471	0.00*
C13	Delay in rectifying defects during service delivery	3.65	7	0.00	3.31	10	0.17	685	−1.411	0.16
C14	Reluctance to seek clarification	2.88	15	0.29	2.94	14	0.69	799.5	−0.322	0.75
C15	Inadequate compensation to displaced persons	3.56	9	0.00	2.66	16	0.08	462	−3.611	0.00*
C16	Unreliable service delivery	3.40	12	0.00	3.50	9	0.00	723	−1.119	0.26

Source: Osei-Kyei et al. (2019a) (*With permission from Elsevier*)
*Significance level 0.05

that it is ranked very high by the Ghanaian respondents. Essentially, Ghana has a multi-party democratic system; therefore, there is often a frequent change of government. Interestingly, when a new government assumes office, it hardly continues the projects/contracts of previous governments particularly if the past government was the opposition political party. This political gymnastic is largely seen in PPP contracts. When a new political party is in administration, it tries to review all past contracts (including PPP arrangements) signed or agreed by the previous government, if it was the opposition party. In the process of reviewing, many contract terms are modified, and this usually infuriates the private partner, which results in long-term dispute with the government or eventually an abrogation of the PPP contract. A typical example is seen in the Build-Own-Operate Transfer (BOOT) Power Plant agreement signed in 2015 between the Government of Ghana and the Africa & Middle East Resources Investment Group LLC (AMERI). In this project, the incumbent government has ordered a review of the deal, with a notion that the contract amount (i.e. US$510 million) was bloated by the previous government, although, the Project Company has asserted that a fair deal was agreed by the previous government. Essentially, the interference of the incumbent government has resulted to misunderstandings and the distress of the project. Unlike Ghana, China is a one-party state, and as one government holds the power and authority of the country, the level of political interference which mostly arises from the change of government as observed in Ghana is very low. Although, there could be some political interference in the Chinese PPP market, it does not certainly lead to the persistent review of contracts and renegotiations which could cause conflicts and distress of PPP projects.

"Delay in decision makings by parties" is ranked first and 11th in Ghana and China respectively. It has a large mean difference of 1.49 (i.e. 4.62 in Ghana and 3.13 in China). Throughout the life-cycle of PPP projects, there are many critical decisions which need to be approved or made by both the public and private sectors to allow the progress of the PPP process. Essentially, lengthy delays in approving or making critical decisions could result to conflicts and disputes (Acharya et al. 2006). In Ghana, the lengthy delay in decision makings in PPPs usually originates from the public sector; and the major cause of the delay is extensive political debate Usually, PPP contracts in Ghana require the assent of Parliament because of the amount involved (MOFEP 2011). Unfortunately, when these PPP contracts are sent to Parliament for approval, they are subjected to lengthy debates and excessive bureaucratic process, which could last for several months. An example was seen in the Ghana National Housing Project, which was initiated in 2009 but unfortunately failed. When this project was sent to Parliament for approval, it was subjected to excessive scrutiny and was rejected by Parliament several times due to technical irregularities. In fact, the lengthy delay in decision making on the part of the government contributed to the failure of the project. Emphatically, when public authorities delay in making decisions or obtaining approvals for PPP contracts, it increases the revenue and financial risks of the private partner, and this could result into misunderstandings and disputes, especially when negotiations have been agreed and require approval from a higher authority. Unlike Ghana, decision-making process in

PPPs in China is not lengthy and complicated. As reported by Ke et al. (2011), public officials in China often quickly finalize negotiations and approvals either based on their personal interests/preferences or career achievement. Further, they asserted that decision makings are easily reached and finalized especially when local government authorities are in dire need of funding for their projects. Nonetheless, because of the political system in China (i.e. centralized government); government officials are able to make decisions and approve PPP contracts without necessarily engaging in political debates or going through tough bureaucratic process. These reasons therefore contribute to the low ranking of this cause of conflict in China compared to Ghana.

"Unrealistic time targets" and "inadequate compensation to displaced persons" are related to poor contractual arrangements. Both causes are ranked higher in Ghana and lower in China. Specifically, "unrealistic time targets" is ranked eighth and 15th in Ghana and China respectively, whereas, "inadequate compensation to displaced persons" is ranked ninth and 16th in Ghana and China respectively. Generally, meeting time targets in PPP arrangements is crucial towards the successful performance of PPP projects and more importantly, it affects revenue collection. Essentially, the earlier the public facility is made available the faster revenue is generated. Due to the lack of experience of many Ghanaian public officials in PPP negotiations and arrangements, unrealistic timelines are often agreed, which in the end are usually not met by the private partners because of many unforeseen circumstances. In addition, because most PPP projects in Ghana are initiated based on campaign promises, very stringent time targets are given to private partners. In the end, when investors are not able to meet these time targets, users and the general public begin to agitate and protest against the government. This situation is same for compensation to displaced persons. As an international best PPP practice, displaced persons should be given compensations for the loss of their properties (particularly land), if it is due to the construction of the PPP project (World Bank 2012). Preferably, compensations have to be equal to the value of the properties of the displaced persons; however, because feasibility studies are not sometimes conducted properly, Ghanaian public officials end up agreeing on compensations which are far below the value of properties of the displaced persons. This therefore triggers public protest and demonstration, which jeopardizes the progress of PPP projects (World Bank 2012; Acharya et al. 2006). In worse cases, the displaced persons engage in lengthy legal battles with the project parties particularly the public authority. Essentially, the Ghanaian situation on the inadequacy of compensation is not different in China. In spite of the fact that China has implemented PPPs over decades, the issue of compensations has still not been properly addressed (Sachs et al. 2007). Zhao (2009) pointed out that since mid-1990s many displaced rural and urban dwellers have strongly protested against the inadequacy of compensations for their properties (particularly farm-lands). However, "inadequate compensation to displaced persons" is ranked lower by the Chinese respondents possibly because in recent years (beginning of 2011), the central government has stepped up the efforts to protect displaced persons by enacting new laws and policies and, establishing monitoring agencies (Zhao 2009; Simon et al. 2016; Faure 2015). These efforts by the Chinese government are helping in minimizing public grievances and disputes over compensations claims in PPPs.

The remaining four causes of conflicts in PPPs with significant differences are ranked higher in China and lower in Ghana. They include unfair risk allocation, unexpected tariff changes, incomplete transfer of risks and absence of proper communication channel. These causes relate to poor risk management and communication. This outcome supports the findings of Chan and Suen (2005), in which improper communication and risk misallocations are the major causes of conflicts in Sino-Foreign Joint Venture projects in China. Unfair risk allocation and incomplete transfer of risks are ranked first and fourth in China respectively, whereas in Ghana, they are positioned 11th and 14th respectively. Many past studies including Ke et al. (2010a, b) and Sachs et al. (2007) have indicated that the misallocation and incomplete transfer of risks are critical in causing contractual disputes in PPPs in China, which often lead to project failure. Sachs et al. (2007) have asserted that many local government authorities hurriedly implement PPP projects, thus they are not able to conduct proper feasibility studies. As a result of this, they end up taking more risks as well as give unrealistic guarantees to the private partner. At the latter stage of the project development, the local government authorities are not able to fulfil their responsibilities considering the costs and this eventually leads to a breach of contract, resulting in lengthy contractual disputes (Ke et al. 2011). This situation is same for incomplete transfer of risks; usually, because feasibility studies are not thorough, extensive and reliable, a comprehensive risk register which explains how risk should be transferred and mitigated is not well developed (Sachs et al. 2007; Ke et al. 2011). It therefore becomes extremely difficult for local authorities and investors to appropriately transfer and mitigate risks, resulting in breach of contracts and potential distress of PPP projects. In Ghana, "unfair risk allocation" and "incomplete transfer of risks" are ranked lower not because contracting authorities are better in PPP risk management; however, many PPP projects have not been implemented as seen in China, in fact, majority of PPP projects are at the feasibility stages. Therefore, conflicts and disputes arising from the poor management PPP risks have not been experienced much as observed in China (Osei-Kyei and Chan 2017). This may have influenced the low rankings of these two causes of conflicts in PPPs.

"Absence of proper communication channel" is ranked third and fifth in China and Ghana respectively. Although, the ranking seems to be quite close, the mean difference is large (i.e. 4.31 and 3.83 for China and Ghana respectively). Chan and Suen (2005) pointed out that the lack of proper communication in relational contracts such as Sino-Foreign Joint Ventures and PPPs in China is mainly due to cultural and language differences. Importantly, a higher proportion of PPP investors in China are foreigners and these foreign investors have little information on the Chinese culture and language (Wang et al. 2012). Many local rules and regulations as well as contract documents are often written in Chinese, which pose serious language barrier (Chan and Suen 2005). Although, some foreign investors engage translators, most of these translators do not have construction background; therefore, they are unable to properly interpret some very technical jargons in the contract documents (Chan and Suen 2005). In addition, most translators do not have in-depth knowledge on the management of PPP arrangements. Because of these, the flow of information among stakeholders (including local residents) is always impeded, resulting in different interpretations, misunderstandings, and eventually a breach of contract (UNDP 2017; Chan and Suen 2005). Unlike China, Ghana follows the western style of procuring PPP projects; more essentially, English is the official language and all contract documents are written in it. In this regard, communication problems mostly resulting from language and cultural differences are not eminent in Ghana's PPP as seen in China.

"Unexpected tariff changes" is ranked fifth in China and 13th in Ghana. Its mean difference is 0.67 (i.e. 3.94 for China and 3.27 for Ghana). In the last few years, unexpected tariff change has become a major problem in the Chinese PPP market particularly in the water sector (Lee 2005). Officially, it is the local authorities who are responsible for setting water tariffs for water PPP projects; however, because of unrealistic guarantees and the lack of accurate estimation of market demand, many water PPP projects end up having high operational and management costs (ADB 2009). In this regard, local authorities usually do not have any option than to increase tariffs to fully cover the water production costs. In fact, what mostly sparks conflicts, particularly between users/public and the project parties is the sudden increment of tariffs (ADB 2009). Tariffs are increased without proper consultation with users; more importantly, there is no transparency in the tariff adjustments (ADB 2009). These therefore enrage the users and tariff adjustments are generally met with negative public reactions. Tariffs are also adjusted for most PPP projects in Ghana particularly concession roads and water projects, however, before tariff adjustments become operative, they have to be approved by the Parliament of Ghana and also scrutinise by other regulatory bodies. Therefore, tariff adjustments are often not sudden and more essentially, they are done with convincing explanations compared to that of China. These may have influenced the low ranking of "unexpected tariff changes" as a cause of conflict in PPPs, particularly between users and project parties in Ghana.

8.5 Conflict Resolution Mechanisms for Public-Private Partnership Projects

Table 8.3 shows the results of the most suitable conflict resolution mechanisms for PPPs in Ghana and China. According to the table, the majority of Ghanaian respondents (i.e. 42.3%) suggest that arbitration should be used in resolving conflicts and disputes in PPPs in Ghana, whereas the majority of respondents (34.4) from China suggest negotiation. This finding is consistent with the outputs of Chan and Suen (2005), in which negotiation was found to be the widely used conflict resolution mechanism in relational contracts such as joint ventures and PPPs in China.

In Ghana, arbitration is generally seen to be more prudent because it has rapid proceedings, allows the project parties to have control over contents and the overall outcome, resolves conflicts in a more private and confidential manner. Furthermore, arbitration ensures that the business relationship between project parties is not lost and damaged (UNDP 2017). In China, Chan and Suen (2005) explained that negotiation is mostly preferred because it is less expensive and preserve the working relationship between the parties involved. They further explained that negotiation is considered by many Chinese as the fairest form of conflict resolution. During negotiation, a 'middle man' is usually engaged to encourage the defaulting party (mostly the local government authority) to honour the terms of contracts or perform its obligations (Chan and Suen 2005).

As also shown in Table 8.3, litigation is indicated by both the Ghanaian and Chinese respondents as the least preferred mechanism for resolving conflicts in PPPs. This outcome is not surprising because litigation drains time and resources. More importantly, it is a win or loss scenario and it damages the relationship between investors and local government authorities (UNDP 2017). In this regard, litigation is often the last resort for resolving conflicts in PPPs (UNDP 2017).

It should be mentioned that the listed preferred conflict resolution mechanisms are used for conflicts occurring between the contracting authority (government) and private partner; as a matter of fact, they are not used to resolve conflicts or disputes between the users/general public and the project parties (i.e. both government and investor). Ideally, the best method of resolving conflicts between users and the project parties is coordination and consultation. In addition, transparency is critical in resolving disputes among users/general and the project parties.

8.6 Conflict Prevention Strategies for Public-Private Partnership Projects

Table 8.4 presents the ranking of 12 conflict prevention strategies for PPP projects. The empirical evidence generated in Table 8.3 is from the Ghanaian respondents only. From the table, it is noticeable that the mean values range between 3.23 and 4.58. This suggests that generally respondents gave high ratings for the list of conflict prevention measures and this demonstrates the overall importance of the

Table 8.3 Most suitable conflict resolution mechanism for PPPs in Ghana and China

	Ghana		China		Ghana and China	
Conflict resolution methods in PPP	Frequency	%	Frequency	%	Frequency	%
Negotiation	19	36.5	11	34.4	30	35.7
Mediation	10	19.2	8	25	18	21.4
Arbitration	22	42.3	7	21.9	29	34.5
Litigation	1	1.9	6	18.8	7	8.3
Total	52	100	32	100	84	100

Source: Osei-Kyei et al. (2019a) (*With permission from Elsevier*)

identified measures. Further, the standard deviation values range from 0.56 to 0.99, which suggests that there is little variation among experts' responses.

To determine the critical conflict prevention measures which need careful consideration by local practitioners, the range normalization method was used. This is whereby prevention measures with normalized values equal to or above 0.50 are considered critical. As presented in Table 8.4, of the 12 conflict prevention measures, four have normalized values above 0.50. In addition, their mean scores range from 3.92 to 4.58. The four measures include "extensive stakeholder consultation in decision makings", "clear goals and mutual benefit objectives", "clarity of roles and responsibilities of parties" and "transparent appeal procedures". This outcome suggests that the four identified critical conflict prevention measures need continuous and more attention by policymakers and investors to avoid potential disputes in PPP arrangement.

The most critical conflict prevention measure is "extensive stakeholder consultation in decision makings". It has a very high mean score of 4.58 and a normalized value of 1.00. This finding is not surprising because many past studies including Osei-Kyei and Chan (2015) and Babatunde et al. (2012) have emphasized the essence of stakeholder consultation in the overall successful performance of PPP projects. However, it should be mentioned that stakeholders in PPP do not only involve the investor and public department/agency but they include external parties such as trade unions, users/commuters and civil group societies (Yuan et al. 2009; Rwelamila et al. 2015). Over the years, the experience of some past PPP projects in developing countries particularly Africa, shows that contracting authorities/public departments usually make critical decisions with respect to the conditions of the partnership and user charges/fees without proper consultation and involvement of the general public and other external parties (Babatunde et al. 2015). This mostly ends up generating conflict between the general public and public officials during the operational period of projects. An example was seen in the Lekki toll road concession project in Nigeria. In the said project, several agitations and

Table 8.4 Mean results of conflict prevention strategies for PPPs

Conflict prevention strategies for PPPs	N	Mean	SD	Normalization
Extensive stakeholder consultation in decision makings	52	4.58	0.75	1.00
Clear goals and mutual benefit objectives	52	4.31	0.78	0.80
Clarity of roles and responsibilities of parties	52	4.27	0.56	0.77
Transparent appeal procedures	52	3.92	0.76	0.51
Appropriate risk sharing	52	3.87	0.71	0.47
Reliable and sustainable service delivery	52	3.81	0.84	0.43
Transparent and user participation in tariffs review	52	3.65	0.81	0.31
Well defined communication systems	52	3.56	0.78	0.24
Realistic conflict risks assessments	52	3.50	0.78	0.20
Quick response to claims	52	3.38	0.99	0.11
Accessible information systems	52	3.29	0.85	0.04
Training and education in communication skills and behaviour	52	3.23	0.90	0.00

Source: Osei-Kyei et al. (2019b) (*With permission from Emerald Publishing Group Ltd*)

demonstrations occurred because of the lack of involvement and consultation of the general public on toll hikes and conditions of the contract (Osei-Kyei and Chan 2015; Amadi et al. 2014). In this regard, the UNDP (2017) suggested that, in developing countries, PPP should be an inclusive arrangement, rather than being exclusive. The UNDP further emphasized that, users and other external parties should be involved from the inception to operation stages of PPP projects. Also, users and other external parties should take center stage in decision makings particularly for pricing, tariff settings and arrangements on service provisions.

"Clear goals and mutual benefit objectives" is the second critical conflict prevention measure for PPPs. It has mean and normalized values of 4.31 and 0.80 respectively. In PPP arrangements both the public and private parties have different and competing motives. For most governments in developing countries, PPPs should be used to promote technology transfer and innovation as well as enhance local economic and social development (Ismail 2013a, b). This is contrary to the priority of private investors, which is to maximize profit and engage in long-term business. The competing priorities are more intense and severe when the PPP arrangement involves a foreign investor, which is often the case in developing countries (Tam 1999). Certainly, if common goal and mutual objectives are not established, it could lead to conflict and dispute in the long term particularly when there is a change in leadership in the public sector. It is therefore important that, during negotiations, both the public and private parties will try as much as to lay down their differences and draw out a common goal and reasonable shared objectives. Further, each party should work toward meeting the defined objectives so that a win-win benefit will be achieved. Importantly, this will reduce the occurrence of conflict and disputes often arising from ambiguous objectives and misplaced priorities.

The third critical prevention measure is "clarity of roles and responsibilities of parties". It has mean and normalized values of 4.27 and 0.77 respectively. This finding is consistent with assertions by the UNDP (2017) that well-defined and agreed duties of project parties minimize the occurrence of conflict in PPPs. Further, Abdul-Aziz and Kassim (2011) mentioned that clearly articulating the responsibilities of parties in a PPP deal contributes substantially to its overall successful performance. Apart from the core responsibilities of the key project parties in PPP arrangements, there are other duties that both parties are sometimes required to perform.

However, if these duties are very unclear and confusing, parties end up not meeting the full expectations of each other and other external parties and this could trigger conflict. For instance, in addition to the primary duties of parties, the investor may be required to provide other social responsibility initiatives such as job training, health-care screening programs and educational outreach for the local community. If the public sector/contracting authority fail to make these roles explicit and seek the full consent of the private sector, some of these initiatives may not be properly implemented. This could generate public outcry concerning the social benefits of the PPP projects. It could even be perceived by some section of the general public that the private investor is gaining as against the host country. Therefore, project parties particularly the contracting authority, should be very clear on all roles and responsibilities, especially for social responsibility initiatives which need to be perform by the private partner. This will help avoid potential conflicts often arising between the general public and the investor at the latter stage of the PPP project.

"Transparent appeal procedures" is the fourth critical conflict prevention measure for PPPs with mean and normalized scores of 3.92 and 0.51, respectively. This finding is also in line with UNDP's (2017) assertion that having a transparent and efficient appeal procedure for PPPs is very instrumental to avoid conflict and disputes in PPP arrangements. Because PPP contracts have complex contractual agreements and responsibilities, at some point, parties may have concerns or complaints which need to be properly addressed. This is often seen in most PPP contracts in developing countries particularly Ghana which involve foreign private investors. In essence, if the existing appeal procedure is poor, inefficient and unfair, it could lead to lengthy contractual disputes and court action. Therefore, it is important for policy makers to establish appeal procedures for PPP arrangements, which are fair and independent (OECD 2008). Moreover, the appeal procedure should ensure that the complainant/accuser is appropriately informed of the progress on any investigations been carried out. It should be mentioned that not only the key project parties should be allowed to lodge complaints or their concerns, users of the facility or general public should also be allowed to appeal when necessary.

8.6.1 Major Groupings of the Conflict Prevention Strategies for Public-Private Partnership Projects

To achieve more interpretable outcome and examine the underlying relationships among the set of 12 conflict prevention measures for PPPs, Factor Analysis (FA) is conducted. Although four factors emerged as significant, it is still important to explore the underlying relationship among the overall set of conflict prevention measures. This will provide more information on how the significant conflict prevention measures should be implemented by local practitioners considering their relationships with other factors.

The factor analysis produced three main factor groupings. These three factor groupings are

1. PG1 – efficient communication structure and risk assessments;
2. PG2 – transparency and openness; and
3. PG3 – proficient service delivery.

8.6.1.1 Efficient Communication Structure and Risk Assessments (PG1)

The sub-factors of this component include "quick response to claims (P10)", "training and education in communication skills and behavior", "realistic conflict risks assessments", "accessible information systems" and "appropriate risk sharing".

Providing training and education to staff in both public and private sectors, accessible information systems and quick response to claims are major steps in ensuring efficient communication structures (Mitkus and Mitkus 2014; UNDP 2017). Normally, because different parties with different ideologies and cultures are engaged in PPPs particularly in developing countries such as Ghana, public officials or personnel need to enhance their communication and relationship skills by undergoing some form of training and education. Essentially, enhancing the communication skills of public officials or personnel will improve their negotiation and relationship skills in PPP deals. Also, to ensure efficient communication structure, project parties should allow the free flow of information and ensure that information is easily accessible by stakeholders (UNDP 2017). Ideally, information on the tendering process, negotiations and project funding could be made available on public institutions' web page. In addition, press conferences could be adopted to improve the accessibility of information on PPP projects by external parties particularly the general public. Providing quick response to claims is also critical in ensuring efficient and effective communication structure. Claims could either be from one of the project parties or external parties. Importantly, whoever needs to respond to claims from any party have to do it within the shortest possible time because lengthy delay could infuriate the complainant and later lead to conflict (Acharya et al. 2006). Ensuring efficiency in risk assessments requires thorough and realistic conflict risk evaluation and the appropriate sharing of project risks among parties. At initial stages of PPP projects, parties should comprehensively review and identify all potential sources of conflicts and quantify the impact of these conflicts on the performance of the project. If parties have very good knowledge on the possible sources and impact of conflict risks, it will help them to devise proper mitigation plans. However, it should be mentioned that conflict risks should be evaluated separately from the overall project risks. Project risks should also be properly identified and allocated to the party with better mitigation plan.

8.6.1.2 Transparency and Openness (PG2)

This factor grouping consists of elements that are much related to creating transparency and openness in PPPs to avoid conflict. This finding is in line with assertions by the UNDP (2017) that transparency and openness in PPPs induces good governance, human rights protection and community-based participation. The sub-factors of this factor component are "extensive stakeholder consultation in decision makings", "transparent appeal procedures", "clear goals and mutual benefit objectives", "transparent and user participation in tariffs review" and "Clarity of roles and responsibilities of parties".

Over the past decades, lack of stakeholder consultations and user participation in critical decision makings in PPPs has resulted to several agitations and disputes among the general public and project parties. Some disputes have led to the complete failure of some projects. In this regard, to ensure adequate openness and transparency, project parties especially the public departments/agencies should engage the general public and other external parties when making critical decisions on tariffs, pricing and conditions of service provisions (UNDP 2017). Moreover, in situations where people/local residents will be relocated because of construction activities, consultations and negotiations should be conducted in advance. Instituting a transparent appeal procedure also contributes substantially in ensuring openness and transparency. Any of the project parties and external stakeholders especially users should be allowed to lodge complaints or express their concerns. The process of petitioning or lodging a complaint has to be independent and fair, and more importantly, the accuser/complainant should be consistently informed of the progress of any investigations being carried out (OECD 2008). Moreover, to enhance transparency and openness it is critical for project parties to establish mutual benefit objectives and explicitly indicate the responsibilities of parties in the partnership. In case one of the project parties is not clear on its duties, enquires should be allowed to be made freely.

8.6.1.3 Proficient Service Delivery (PG3)

This factor structure consists of two sub-factors namely, "reliable and sustainable service delivery" and "well defined communication systems". A well-defined communication system allows service operators to receive useful feedback and concerns from customers/users which enables them to improve their service delivery. In fact, it will be extremely difficult for service operators to meet the expectation of users/general public and provide efficient services, if feedbacks are not allowed from customers through proper communication system. Providing uninterrupted services is a good mark of proficient service delivery. However, for service operators to provide uninterrupted services, they need to adopt planned preventive maintenance instead of corrective maintenance.

8.7 Chapter Summary

This chapter has comprehensively discussed the management of conflicts by focusing on the root causes, resolution mechanisms and prevention strategies from Chinese and Ghanaian perspectives. From the discussion, it was realised that the causes of conflict ranked higher in Ghana directly relate to poor governance and contract arrangement, whereas causes related to poor risk management and communication are ranked higher in China. Further, arbitration and negotiation are the most suitable conflict resolution mechanisms for PPPs in Ghana and China respectively. Lastly, the most critical conflict prevention strategies for PPP are "extensive stakeholder consultation in decision makings", "clear goals and mutual benefit objectives", "clarity of roles and responsibilities of parties" and "transparent appeal procedures".

References

Abdul-Aziz, A. R., & Jahn Kassim, P. S. (2011). Objectives, success and failure factors of housing public–private partnerships in Malaysia. *Habitat International, 35*(1), 150–157.

Acharya, N. K., Dai Lee, Y., & Man Im, H. (2006). Conflicting factors in construction projects: Korean perspective. *Engineering, Construction and Architectural Management, 13*(6), 543–566.

Amadi, C., Carrillo, P., & Tuuli, M. (2014). Stakeholder management in public private partnership projects in Nigeria: Towards a research agenda. In A. B. Raiden & E. Aboagye-Nimo (Eds.), *Proceedings of the 30th annual ARCOM conference, 1–3 September 2014, Portsmouth, UK.*

Asian Development Bank (ADB). (2009). *Understanding urban water tariff in the People's Republic of China.* Asian Development Bank.

Babatunde, S. O., Opawole, A., & Akinsiku, O. E. (2012). Critical success factors in public-private partnership (PPP) on infrastructure delivery in Nigeria. *Journal of Facilities Management, 10*(3), 212–225.

Babatunde, S. O., Perera, S., Zhou, L., & Udeaja, C. (2015). Barriers to public private partnership projects in developing countries: A case of Nigeria. *Engineering, Construction and Architectural Management, 22*(6), 669–691.

Chan, E. H., & Suen, H. C. (2005). Dispute resolution management for international construction projects in China. *Management Decision, 43*(4), 589–602.

Cheung, E., Chan, A. P. C., & Kajewski, S. (2012a). Factors contributing to successful public private partnership projects, Comparing Hong Kong with Australia and the United Kingdom. *Journal of Facilities Management, 10*(1), 45–58.

Cheung, E., Chan, A. P. C., Lam, P. T. I., Chan, D. W. M., & Ke, Y. (2012b). A comparative study of critical success factors for public private partnerships (PPP) between Mainland China and the Hong Kong Special Administrative Region. *Facilities, 30*(13/14), 647–666.

Ejohwomu, A. O., Oshodi, O. S., & Onifade, M. K. (2016). Causes of conflicts in construction projects in Nigeria: Consultants and contractors perspective. *Nigerian Journal of Technology, 35*(2), 270–277.

Faure, A. (2015). *Public international law debate concerning forced evictions in China.* Browser Download This Paper.

Gardiner, P. D., & Simmons, J. E. (1992). Analysis of Conflict and Change in Construction Projects. *Construction Management and Economics, 10*(6), 459–478.

Gorse, C. A. (2003). Conflict and conflict management in construction. In *Proceedings of 19th annual conference of the Association of Researchers in Construction Management* (pp. 173–182).

Harmon, K. M. (2003). Conflicts between owner and contractors: proposed intervention process. *Journal of Management in Engineering, 19*(3), 121–125.

Ismail, S. (2013a). Critical success factors of public private partnership (PPP) implementation in Malaysia. *Asia-Pacific Journal of Business Administration, 5*(1), 6–19.

Ismail, S. (2013b). Positive and negative factors for the successful adoption of Public Private Partnership: evidence from Malaysia. In *Public Private Partnerships (PPP): Malaysian studies* (pp. 65–81). Kuala Lumpur: IIUM Press.

Jaffar, N., Tharim, A. A., & Shuib, M. N. (2011). Factors of conflict in construction industry: a literature review. *Procedia Engineering, 20*, 193–202.

Ke, Y., Wang, S., & Chan, A. P. C. (2010a). Risk allocation in public-private partnership infrastructure projects: Comparative study. *Journal of Infrastructure Systems, 16*, 343.
Ke, Y., Wang, S., Chan, A. P., & Lam, P. T. (2010b). Preferred risk allocation in China's public–private partnership (PPP) projects. *International Journal of Project Management, 28*(5), 482–492.
Ke, Y., Wang, S., Chan, A. P., & Cheung, E. (2011). Understanding the risks in China's PPP projects: ranking of their probability and consequence. *Engineering, Construction and Architectural Management, 18*(5), 481–496.
Kumaraswamy, M. M. (1997). Conflicts, claims and disputes in construction. *Engineering, Construction and Architectural Management, 4*(2), 95–111.
Lee, V. (2005). *Public – Private Partnerships*. Hong Kong: *Research and Library Services Division Legislative Council Secretariat.*
Ministry of Finance and Economic Planning. (2011). *National policy on public – private partnerships*. Ghana.
Mitkus, S., & Mitkus, T. (2014). Causes of conflicts in a construction industry: A communicational approach. *Procedia-Social and Behavioral Sciences, 110*, 777–786.
Organisation for Economic Co-operation and Development. (2008). *Public-private partnerships: In pursuit of risk sharing and value for money*. Paris: OECD Publications.
Osei-Kyei, R., & Chan, A. P. C. (2015). Developing transport infrastructure in Sub-Saharan Africa through public–private partnerships: Policy practice and implications. *Transport Reviews*, 1–17.
Osei-Kyei, R., & Chan, A. P. (2017). Risk assessment in public-private partnership infrastructure projects: empirical comparison between Ghana and Hong Kong. *Construction Innovation*, 204–223.
Osei-Kyei, R., Chan, A. P., Dansoh, A., Ofori-Kuragu, J. K., & Oppong, G. D. (2018a). Strategies for effective management of unsolicited public–private partnership proposals. *Journal of Management in Engineering, 34*(3), 04018006.
Osei-Kyei, R., Chan, A. P., Dansoh, A., Ofori-Kuragu, J. K., & Owusu, E. K. (2018b). Motivations for adopting unsolicited proposals for public-private partnership project implementation: a survey of international experts. *Journal of Financial Management of Property and Construction*, 221–238.
Osei-Kyei, R., Chan, A. P., Yu, Y., Chen, C., & Dansoh, A. (2019a). Root causes of conflict and conflict resolution mechanisms in public-private partnerships: Comparative study between Ghana and China. *Cities, 87*, 185–195.
Osei-Kyei, R., Chan, A. P., Yao, Y., & Mazher, K. M. (2019b). Conflict prevention measures for public-private partnerships in developing countries. *Journal of Financial Management of Property and Construction*, 39–57.
Rwelamila, P. D., Fewings, P., & Henjewele, C. (2015). Addressing the missing link in PPP projects: What constitutes the public? *Journal of Management in Engineering, 31*(5), 04014085.
Sachs, T., Tiong, R., & Qing Wang, S. (2007). Analysis of political risks and opportunities in public private partnerships (PPP) in China and selected Asian countries: Survey results. *Chinese Management Studies, 1*(2), 126–148.
Simon, T. W., Feng, C., Nelson, L. P. (2016). *China's changing legal system Lawyers & Judges on Civil & Criminal Law*. US: Palgrave Macmillan.
Tam, C. M. (1999). Build-operate-transfer model for infrastructure developments in Asia: Reasons for successes and failures. *International Journal of Project Management, 17*(6), 377–382.
United Nations Development Programme (UNDP). (2017). *Implementation – Managing conflict: What are conflicts*, UNDP. Available from: http://pppue.undp.2margraf.com/en/20_3.htm. Accessed on Feb 2017.
Wang, S. Q., Ke, Y. J., & Xie, J. (2012). *Public private partnership implementation in China*, in G.M.
World Bank. (2012). *Ghana – Public Private Partnership Project (PPP)*. Washington, DC: World Bank.
World Bank. (2015a). *Private Participation in Infrastructure (PPI), regional snapshots*. Washington, DC: World Bank. Available from: http://ppi.worldbank.org/explore/ppi_exploreRegion.aspx?regionID=1. Accessed on May 2015.

World Bank. (2015b). *Private Participation in Infrastructure (PPI), Middle East and North Africa. Regional snapshots*. Washington, DC: World Bank. Available from: http://ppi.worldbank.org/explore/ppi_exploreRegion.aspx?regionID=1. Accessed on May 2015.

World Bank. (2015c). *Data: Population total*. Available from: http://data.worldbank.org/country

World Bank. (2017). *Guidelines for the development of a policy for man-aging unsolicited proposals in infrastructure projects*. Washington, DC. http://pubdocs.worldbank.org/en/976121488983070966/Guidelines-for-the-Development-of-a-USP-Policy.pdf. May 2017.

Yuan, J., Zeng, A. Y., Skibniewski, M. J., & Li, Q. (2009). Selection of performance objectives and key performance Indicators in public-private partnerships projects to achieve value for money. *Construction Management and Economics, 27*(3), 253–270.

Zhao, B. (2009). Land expropriation, protest, and impunity in rural China. *Focaal, 2009*(54).

Chapter 9
A Best Practice Framework for Public-Private Partnership

Abstract This chapter presents a practice framework for implementing successful PPP projects. The practice framework consists of a checklist of best practices which both contracting authorities and private sector partners should consider and adhere to in order to achieve successful PPP projects. Emphatically, the best practices proposed are based on the empirical findings reported in the previous chapters. The best practices are then incorporated in the PPP process.

Keywords Best practice framework · PPP process · Project inception · Procurement stage · Construction stage · Operation and transfer stage

9.1 Introduction

This chapter presents a practice framework for implementing successful PPP projects. The practice framework consists of a checklist of best practices which both contracting authorities and private sector partners should consider and adhere to in order to achieve successful PPP projects. Emphatically, the best practices proposed are based on the empirical findings reported in the previous chapters. The best practices are then incorporated in the PPP process.

R. Osei-Kyei, A. P. C. Chan, *International Best Practices of Public-Private Partnership*, https://doi.org/10.1007/978-981-33-6268-0_9

9.2 Best Practice Framework for PPP Implementation (Figs. 9.1, 9.2, 9.3, and 9.4)

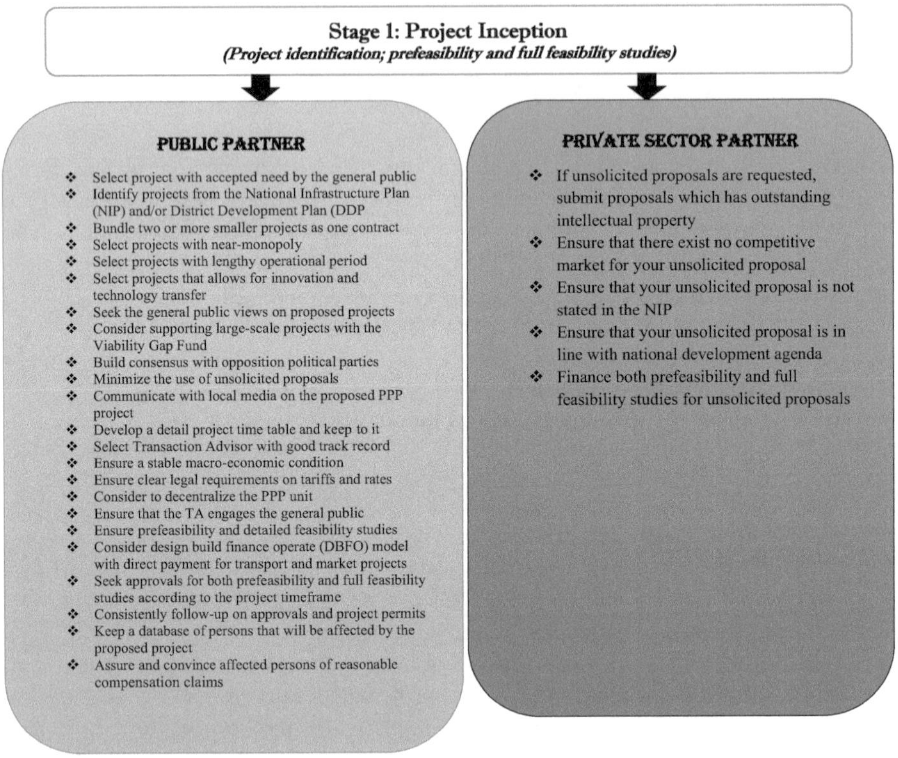

Fig. 9.1 Best practices for Project Inception stage of PPP process. (Osei-Kyei and Chan 2018, *with permission from Emerald*)

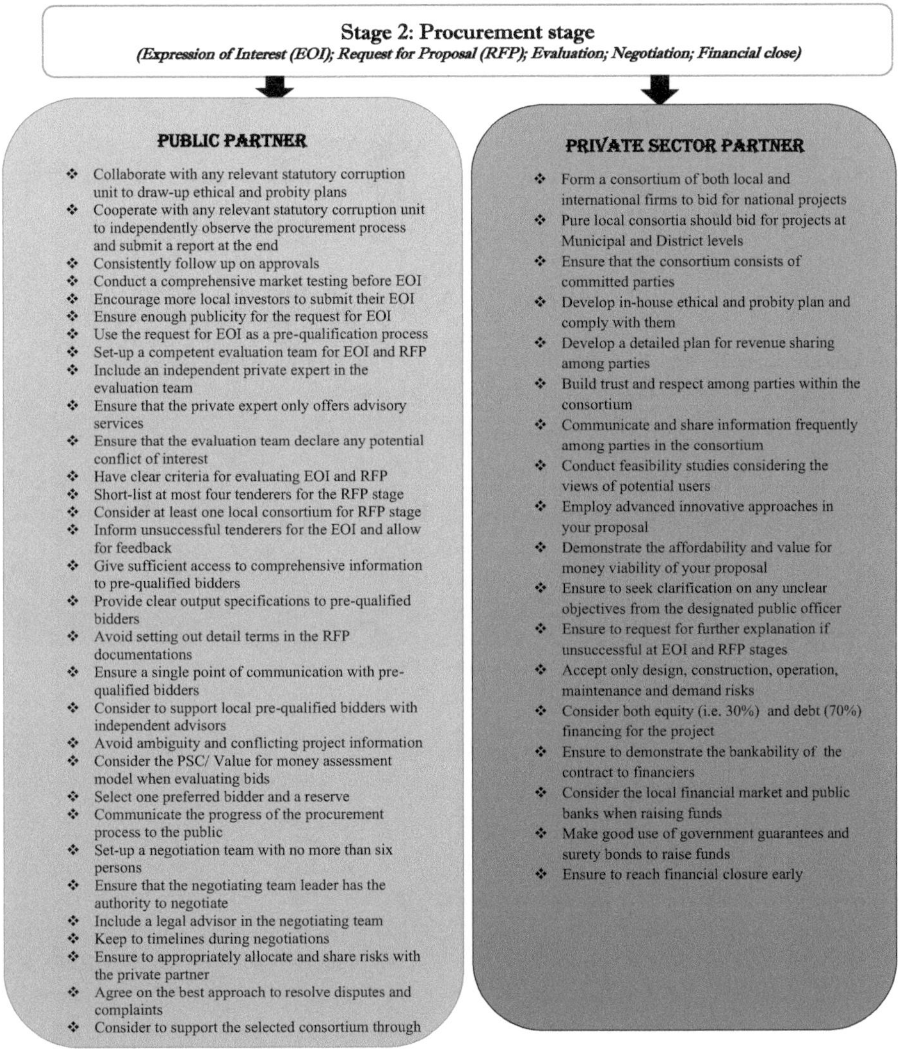

Fig. 9.2 Best practices for Procurement stage of PPP process. (Osei-Kyei and Chan 2018, *with permission from Emerald*)

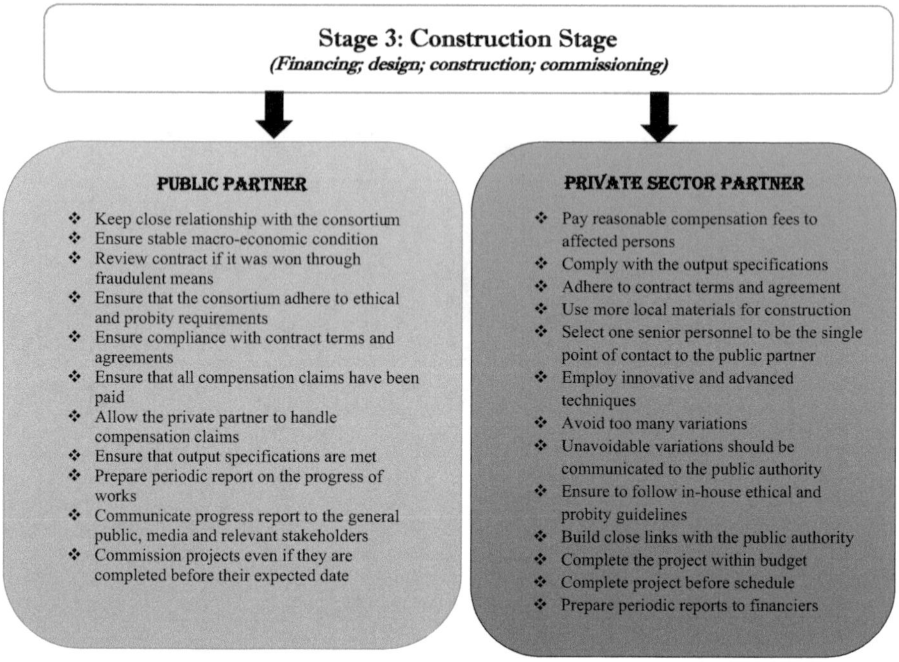

Fig. 9.3 Best practices for Construction stage of PPP process. (Osei-Kyei and Chan 2018, *with permission from Emerald*)

9.3 Chapter Summary

This chapter has presented a best practice framework for PPP implementation. The framework was developed based on research findings reported in previous chapters. The best practices are categorized according to the PPP process. This best practice framework will inform policy makers and private sector partners of the strategies and measures that should be carefully observed at each stage of the PPP process.

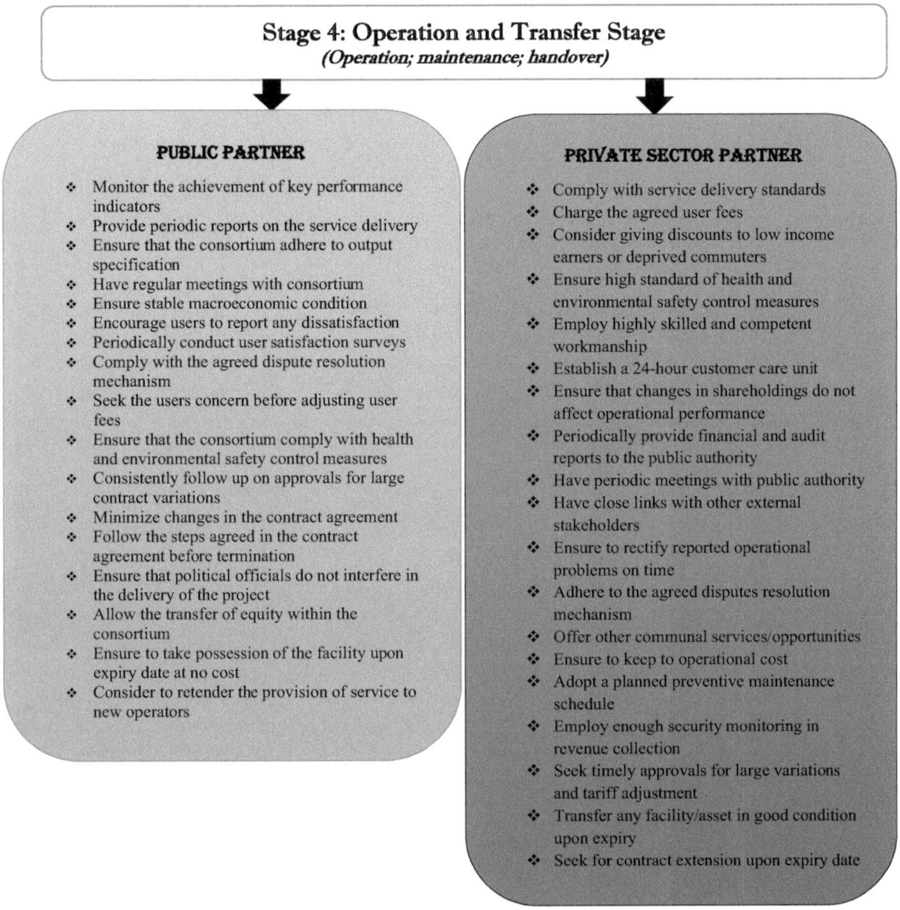

Fig. 9.4 Best practices for Operational stage of PPP process. (Osei-Kyei and Chan 2018, *with permission from Emerald*)

Reference

Osei-Kyei, R., & Chan, A. P. (2018). A best practice framework for public-private partnership implementation for construction projects in developing countries. *Benchmarking: An International Journal*.

Bibliography

Abadie, R., & Howcroft, A. (2004). *Developing PPPs in New Europe*. London: Price Waterhouse Coopers.

Abdel Aziz, A. M. (2007). Successful delivery of public-private partnerships for infrastructure development. *Journal of Construction Engineering and Management, 133*(12), 918–931.

Abdel Aziz, A., & Nabavi, H. (2014). Unsolicited proposals for PPP projects: Private sector perceptions in the USA. In *Construction Research Congress 2014: Construction in a global network* (pp. 1349–1358). Reston: American Society of Civil Engineers.

Abdi, H. (2003). *Factor rotations in factor analyses. Encyclopaedia for research methods for the social sciences*. Thousand Oaks: Sage.

Abdul-Aziz, A. R., & Jahn Kassim, P. S. (2011). Objectives, success and failure factors of housing public–private partnerships in Malaysia. *Habitat International, 35*(1), 150–157.

Abramson, H. N., Encarnacao, J., Reid, P. P., & Schmoch, U. (Eds.). (1997). *Technology transfer systems in the United States and Germany: lessons and perspectives*. Washington, DC: National Academies Press.

Acharya, N. K., Dai Lee, Y., & Man Im, H. (2006). Conflicting factors in construction projects: Korean perspective. *Engineering, Construction and Architectural Management, 13*(6), 543–566.

Africa PPP. (2014). *PPP conference communique*. Available from (Aug. 2014); www.africappp.com/wp.../APPP_communique_PPP-Country-Focus.pdf

African Development Bank Group (ADBG). (2008). *Nigeria: AfDB Approves US$ 85 Million for Lekki Toll Road Project*. Available from: http://www.afdb.org/en/news-and-events/article/nigeria-afdb-approves-us-85-million-for-lekki-toll-road. Accessed on 31 Dec 2014.

African Elections Project. (2010). *Minister outdoors plan for STX Project*. Available from: http://www.africanelections.org//ghana/postelections/news/page.php?news=451. Accessed on 7 Nov 2014.

Ahadzie, D. K., Proverbs, D. G., & Olomolaiye, P. O. (2008). Critical success criteria for mass house building projects in developing countries. *International Journal of Project Management, 26*, 675–687.

Akanbang, B. A. A. (2010), "*Public-Private Partnerships in Small Town Water Supply in Tumu", Tripartite Partnership Project, Ghana*. Available from: www.ghana.watsan.net/page/777 (Aug. 2014).

Akintoye, A., & Beck, M. (2009). *Policy, management and finance for public-private partnerships*. West Sussex: Wiley.

R. Osei-Kyei, A. P. C. Chan, *International Best Practices of Public-Private Partnership*, https://doi.org/10.1007/978-981-33-6268-0

Akintoye, A., Hardcastle, C., Beck, M., Chinyio, E., & Asenova, D. (2003). Achieving best value in private finance initiative project procurement. *Construction Management and Economics, 21*(5), 461–470.

Akintoye, A., Beck, M., & Hardcastle, C. (2008). *Public-private partnerships: Managing risks and opportunities*. Blackwell Science.

Akintoye, A., Beck, M., & Kumaraswamy, M. (Eds.). (2015). *Public private partnerships: A global review*. Routledge.

Algarni, A. M., Arditi, D., & Polat, G. (2007). Build-operate-transfer in infrastructure projects in the United States. *Journal of Construction Engineering and Management, 133*(10), 728–735.

Al-Tmeemy, S. M. H. M., Abdul-Rahman, H., & Harun, Z. (2011). Future criteria for success of building projects in Malaysia. *International Journal of Project Management., 29*(3), 337–348.

Amadi, C., Carrillo, P., & Tuuli, M. (2014). Stakeholder management in public private partnership projects in Nigeria: Towards a research agenda. In A. B. Raiden & E. Aboagye-Nimo (Eds.), *Proceedings of the 30th annual ARCOM conference, 1–3 September 2014, Portsmouth, UK*.

Amesimeku E. (2013). *Ghanaians express cautious optimism on new affordable housing project. Property Investor Africa*, 23 December. Available from: https://propertyinvestorafrica.com/ghanaians-express-cautious-optimism-on-new-affordable-housing-project. Accessed on 1 Nov 2014.

Ameyaw E.E (2015). *Risk allocation model for Public-Private Partnership water supply projects in Ghana* (Doctoral dissertation, The Hong Kong Polytechnic University).

Ameyaw, E. E., & Chan, A. P. C. (2013). Identifying public-private partnership (PPP) risks in managing water supply projects in Ghana. *Journal of Facilities Management, 11*(2), 152–182.

Ameyaw, E. E., & Chan, A. P. (2015a). Risk allocation in public-private partnership water supply projects in Ghana. *Construction Management and Economics, 33*(3), 187–208.

Ameyaw, E. E., & Chan, A. P. (2015b). Evaluation and ranking of risk factors in public–private partnership water supply projects in developing countries using fuzzy synthetic evaluation approach. *Expert Systems with Applications, 42*(12), 5102–5116.

Ameyaw, E. E., & Chan, A. P. C. (2015c). Risk ranking and analysis in PPP water supply infrastructure projects: an international survey of industry experts. *Facilities, 33*(7/8), 428–453.

Ameyaw, E. E., Chan, A. P., Owusu-Manu, D. G., & Coleman, E. (2015). A fuzzy model for evaluating risk impacts on variability between contract sum and final account in government-funded construction projects. *Journal of Facilities Management, 13*(1), 45–69.

Angeriz, A., & Arestis, P. (2007). Monetary policy in the UK. *Cambridge Journal of Economics, 31*(6), 863–884.

Annual Report. (2008). *Inkosi Albert Luthuli Central Hospital*. Available from: http://www.ialch.co.za/index.php/resources/annual-reports?download=3:annual-report-2007-2008. Accessed Feb 2015.

Appleby, G., Aroney, N., & John, T. (2012). *The future of Australian federalism: comparative and interdisciplinary perspectives*. Cambridge University Press.

Arndt, R.H. (2000). *Getting a fair deal: Efficient risk allocation in the private provision of infrastructure*. PhD thesis, Department of Civil and Environmental Engineering, the University of Melbourne.

Asare, B. E., & Frimpong, M. K. (2013). Public-private partnerships and urban sanitation: Do expectations meet realities in Madina-Ghana? *Journal of African Studies and Development, 5*(5), 113–124.

Ashley, D. B., Lurie, C. S., & Jaselskis, E. J. (1987). Determinants of construction project success. *Project Management Journal, 18*(2), 69–79.

Asia World Expo. (2010). *Asia World-Expo continues to add tremendous value to Hong Kong's exhibition industry and economy*. Available from http://www.asiaworld expo.com.hk/html/en/NewsRoom/PressReleaseDetail_ff73a51fda6e4a7383c60bf523a605bf.html. Accessed on May 2015.

Asia World Expo. (2014a). *Floor plan*. Available from http://www.asiaworld-expo.com.hk/html/en/NewsRoom/FloorPlan.html. Accessed on May 2015.

Asia World Expo. (2014b). *Fact sheet*. Available from http://www.asiaworld-expo.com.hk/html/en/NewsRoom/FactSheet.html. Accessed on May 2015.

Asian Development Bank (ADB). (2009). *Understanding urban water tariff in the People's Republic of China*. Asian Development Bank.

Asian Development Bank (ADB). (2011). *Public-private partnership infrastructure projects: Case studies from the Republic of Korea*. Available from: http://ppp.worldbank.org/public-private-partnership/library/public-private-partnership-infrastructure-projects-case-studies-republic-korea-0. Accessed on May 2015.

Askar, M., & Gab-Allah, A. (2002). Problems facing parties involved in build operate, and transport projects in Egypt. *Journal of Management in Engineering, 18*(4), 173–178.

Atkinson, R. (1999). Project management: cost, time and quality, two best guesses and a phenomenon, it's time to accept other success criteria. *International Journal of Project Management, 17*(6), 337–342.

Aurecon. (2014). *Lekki-Epe Expressway Toll Road, Nigeria. Aurecon Group*. Available from: http://www.aurecongroup.com/en/projects/transport/lekki-epe-expressway-toll-road. Accessed on 4 Jan 2015.

Awortwi, N. (2004). Getting the fundamentals wrong: woes of public–private partnerships in solid waste collection in three Ghanaian cities. *Public Administration and Development, 24*(3), 213–224.

Babatunde, S. O., Opawole, A., & Akinsiku, O. E. (2012). Critical success factors in public-private partnership (PPP) on infrastructure delivery in Nigeria. *Journal of Facilities Management, 10*(3), 212–225.

Babatunde, S. O., Perera, S., Zhou, L., & Udeaja, C. (2015). Barriers to public private partnership projects in developing countries: A case of Nigeria. *Engineering, Construction and Architectural Management, 22*(6), 669–691.

Baccarini, D. (1999). The logical framework method for defining project success. *Project Management Journal, 30*, 25–32.

Bajaja, D., Oluwoye, J., & Lenard, D. (1997). An Analysis of Contractor's approaches to Risk Identification in New South Wales, Australia. *Construction Management and Economics, 15*, 363–369.

Ballingall, J. (2014). *PPPs: Fiscal space and investment: The United Kingdom*. Infrastructure U.K. http://idbdocs.iadb.org/wsdocs/getdocument.aspx?docnum=38831864i. Mar 2017.

Beauregard, R. A. (1998). Public-private partnerships as historical chameleons: the case of the United States. In J. Pirre (Ed.), *Partnerships in urban governance*. London: MacMillan.

Benson I. (2011). STX Deal to start at last. *The Chronicle*, November 16. Available from: http://thechronicle.com.gh/stx-deal-to-start-at-last/. Accessed on 23 Oct 2014.

Berkoh, C., Hirsch, D., Larbi, E., Simavi, F. L., Ntow, S., Smet, J., & Simavi, M. T. (2004). *Report on the feasibility of public-private partnership for sustainable water supply to the urban poor in Ghana*. Delft: Netherlands Water Partnership.

Bhatia, B., & Gupta, N. (2006). *Lifting constraints to public-private partnerships in South Asia*. Washington, DC: World Bank. © World Bank.

Boafo-Arthur, K. (Ed.). (2007). *Ghana: One decade of the liberal state*. New York: Zed Books.

Boussabaine, A. (2013). *Cost planning of PFI and PPP building projects*. London: Routledge.

Bovis, C. (2013). *Public-private Partnerships in the European Union*. London: Routledge.

Boylan, H. (2012). *Public–private partnerships in Africa, Part I – Infrastructure*. Consultancy Africa Intelligence. Retrieved January 20, 2015, from www.consultancyafrica.com

Broadbent, J., & Laughlin, R. (2003). Public private partnerships: an introduction. *Accounting, Auditing & Accountability Journal, 16*(3), 332–341.

Brown, A., & Adams, J. (2000). Measuring the effect of project management on construction outputs: a new approach. *International Journal of Project Management, 18*, 327–335.

Bryde, D. J., & Robinson, L. (2005). Client versus contractor perspectives on project success criteria. *International Journal of Project Management, 23*, 622–629.

Buxbaum, J. N., & Ortiz, I. N. (2009). *Public sector decision making for public-private partnerships* (Vol. 391). Transportation Research Board. U.S.A.

Canadian Council for Public Private Partnerships (CCPPP). (2013). *10 – year economic impact assessment of PPP in Canada 2003–2012. PPP Canada.* Available from: http://www.p3canada.ca/en/about-p3s/p3-resource-library/10-year-economic-impact-assessment-of-public-private-partnerships-in-canada. Accessed May 2015.

Carroll, P., & Steane, P. (2000). Public-private partnerships: Sectoral perspectives. *Routledge Advances in Management and Business Studies, 19*, 36–56.

Chan, A. P. C., & Ameyaw, E. E. (2013). "The private sector's involvement in the water industry of Ghana", *Journal of Engineering. Design and Technology, 11*(3), 251–275.

Chan, A. P. C., & Chan, A. P. (2004). Key performance indicators for measuring construction success. *Benchmarking: An International Journal, 11*(2), 203–221.

Chan, A. P. C., & Cheung, E. (2014). *Public private partnerships in international construction: Learning from case studies*. New York: Routledge.

Chan, D. W., & Kumaraswamy, M. M. (1999). Forecasting construction durations for public housing projects: A Hong Kong perspective. *Building and Environment, 34*(5), 633–646.

Chan, E. H., & Suen, H. C. (2005). Dispute resolution management for international construction projects in China. *Management Decision, 43*(4), 589–602.

Chan, A. P., Scott, D., & Lam, E. W. (2002). Framework of success criteria for design/build projects. *Journal of Management in Engineering*. https://doi.org/10.1061/(ASCE)0742-597X(2002)18:3(120).

Chan, A. P., Chan, D. W., Chiang, Y. H., Tang, B. S., Chan, E. H., & Ho, K. S. (2004). Exploring critical success factors for partnering in construction projects. *Journal of Construction Engineering and Management, 130*(2), 188–198.

Chan, D. W. M., Chan, A. P. C., & Lam, P. T. I. (2006). *A feasibility study of the implementation of public private partnership (PPP) in Hong Kong.* Proceedings of the CIB W89. International conference on building education and research, Hong Kong, April 10–13, 2006.

Chan, A. P. C., Sidwell, T., Kajewski, S., Lam, P. T. I, Chan, D. W. M., & Cheung, E. (2007). *From BOT to PPP – A Hong Kong example.* Proceedings of the 2007 International Conference on Concession Public/Infrastructural Projects (ICCPIP), Vol. 9, Dalian University of Technology, China, pp. 010–018.

Chan, A. P. C., Lam, P. T., Chan, D. W., Cheung, E., & Ke, Y. (2009). Drivers for adopting public private partnerships—empirical comparison between China and Hong Kong special administrative region. *Journal of Construction Engineering and Management, 135*(11), 1115–1124.

Chan, A. P., Lam, P. T., Chan, D. W., Cheung, E., & Ke, Y. (2010a). Potential obstacles to successful implementation of public-private partnerships in Beijing and the Hong Kong special administrative region. *Journal of Management in Engineering*. https://doi.org/10.1061/(ASCE)0742-597X(2010)26:1(30).

Chan, A. P., Lam, P. T., Chan, D. W., Cheung, E., & Ke, Y. (2010b). Critical success factors for PPPs in infrastructure developments: Chinese perspective. *Journal of Construction Engineering and Management, 136*(5), 484–494.

Chan, A. P. C., Yeung, J. F., Calvin, C. P., Wang, S. Q., & Ke, Y. (2011). Empirical study of risk assessment and allocation of public-private partnership projects in China. *Journal of Management in Engineering, 27*(3), 136–148.

Chan, A. P., Lam, P. T., Wen, Y., Ameyaw, E. E., Wang, S., & Ke, Y. (2014). Cross-sectional analysis of critical risk factors for PPP water projects in China. *Journal of Infrastructure Systems, 21*(1), 04014031.

Cheung, E. (2008). Developing a best practice framework for implementing public private partnerships (PPP) in Hong Kong. PhD Thesis. Queensland University of Technology, Australia.

Cheung, E. (2009). *Developing a best practice framework for implementing public private partnerships (PPP) in Hong Kong* (Doctoral dissertation, Queensland University of Technology).

Cheung, E., & Chan, A. P. C. (2009). Is BOT the best financing model to procure infrastructure projects?: A case study of the Hong Kong-Zhuhai-Macau Bridge. *Journal of Property Investment & Finance, 27*(3), 290–302.

Cheung, E., & Chan, A. P. C. (2011a). Evaluation model for assessing the suitability of public-private partnership projects. *Journal of Management in Engineering*. https://doi.org/10.1061/(ASCE)ME.1943-5479.0000044.

Cheung, E., & Chan, A. (2011b). Risk factors of public-private partnership projects in China: Comparison between the water, power, and transportation sectors. *Journal of Urban Planning and Development, 137*(4), 409–415.

Cheung, S. O., Tam, C. M., Ndekugri, I., & Harris, F. C. (2000). Factors affecting client's project dispute resolution satisfaction in Hong Kong. *Construction Management Economic, 18*(3), 281–294.

Cheung, E., Chan, A. P., & Kajewski, S. (2009). Reasons for implementing public private partnership projects: Perspectives from Hong Kong, Australian and British practitioners. *Journal of Property Investment & Finance, 27*(1), 81–95.

Cheung, E., Chan, P. C., & Kajewski, S. (2010a). Suitability of procuring large public works by PPP in Hong Kong. *Engineering, Construction and Architectural Management, 17*(3), 292–308.

Cheung, E., Chan, A. P., & Kajewski, S. (2010b). The public sector's perspective on procuring public works projects-comparing the views of practitioners in Hong Kong and Australia. *Journal of Civil Engineering and Management, 16*(1), 19–32.

Cheung, E., Chan, A. P. C., & Kajewski, S. (2012a). Factors contributing to successful public private partnership projects, Comparing Hong Kong with Australia and the United Kingdom. *Journal of Facilities Management, 10*(1), 45–58.

Cheung, E., Chan, A. P. C., Lam, P. T. I., Chan, D. W. M., & Ke, Y. (2012b). A comparative study of critical success factors for public private partnerships (PPP) between Mainland China and the Hong Kong Special Administrative Region. *Facilities, 30*(13/14), 647–666.

Chou, J. S., Tserng, H. P., Lin, C., & Yeh, C. P. (2012). Critical factors and risk allocation for PPP policy: Comparison between HSR and general infrastructure projects. *Transport Policy, 22*, 36–48.

Chow, L. K. (2005). *Incorporating fuzzy membership functions and gap analysis concept into performance evaluation of engineering consultants: Hong Kong study* (Doctoral dissertation, The University of Hong Kong, Pokfulam, Hong Kong).

Cisse, H., Menon, N. M., Segger, M. C. C., & Nmehielle, V. O. (2013). *The World Bank legal review. Volume 5: Fostering development through opportunity, inclusion, and equity*. Washington, DC: World Bank Publications.

Citifmonline. (2011). *BK Asamoah: I'm still STX Ghana Boss*. Available from: http://www.ghanaweb.com/GhanaHomePage/NewsArchive/artikel.php?ID=216675. Accessed on 2 Oct 2014.

Coeuré, B., & Jacquet, P. (2010). *Economic policy: Theory and practice*. New York: Oxford University Press.

Cohen, D. J., & Crabtree, B. F. (2008). Evaluative criteria for qualitative research in health care: controversies and recommendations. *The Annals of Family Medicine, 6*(4), 331–339.

Collin, S. (1998). In the twilight zone: a survey of public-private partnerships in Sweden. *Public Productivity & Management Review, 21*(3), 272–283. San Francisco: Sage.

Commission of the European Communities. (2000). *Proposal for a Directive of European Parliament and The Council on coordination of procedures for the award of public supply contracts, public service contracts.*

Corbett, P., & Smith, R. (2006). *An analysis of the success of the private finance initiative as the government's preferred procurement route*. Proceedings of the Accelerating Excellence in the Built Environment Conference, Birmingham, UK, October 2–4.

Corrigan, M. B., Hambene, J., Hudnut, W., III, Levitt, R. L., Stainback, J., Ward, R., & Witenstein, N. (2005). *Ten principles for successful public/private partnerships*. Washinton, DC: Urban Land Institute.

Coulson, A. (2005). A plague on all your partnerships: theory and practice in regeneration. *International Journal of Public Sector Management, 18*(2), 151–163.

Cox, R. F., Issa, R. R. A., & Ahrens, D. (2003). Management's perception of key performance indicators for construction. *Journal of Construction Engineering and Management, 129*(2), 142–151.

Cserháti, G., & Szabó, L. (2014). The relationship between success criteria and success factors in organisational event projects. *International Journal of Project Management, 32*(4), 613–624.

Cuttaree, V., & Mandri-Perrott, X. C. (2011). *Public–private partnerships in Europe and Central Asia: Designing crisis-resilient strategies and bankable projects*. Washington, DC: World Bank Publications.

Daily Guide. (2010). *GREDA chickens out of opposition to STX housing deal*. Available from: http://www.modernghana.com/news/283489/1/greda-chickens-out-of-opposition-to-stx-housing-de.html. Accessed on 16 Nov 2014.

Demuijnck, G., & Ngnodjom, H. (2011). Public-private partnerships and corruption in developing countries: A case study. *Business & Professional Ethics Journal – Center for Applied Philosophy, 3*, 253–269.

Department of the Environment, Heritage and Local Government. (2003). *Policy framework for public private partnership (PPP) in Ireland*. Ireland: Irish Government.

Dixon, T., Pottinger, G., & Jordan, A. (2005). Lessons from the private finance initiative in the UK: Benefits, problems and critical success factors. *Journal of Property Investment & Finance, 23*(5), 412–423.

Doloi, H. (2012). Understanding impacts of time and cost related construction risks on operational performance of PPP projects. *International Journal of Strategic Property Management, 16*(3), 316–337.

Dulaimi, M. F., Alhashemi, M., Ling, F. Y. Y., & Kumaraswamy, M. (2010). The execution of public–private partnership projects in the UAE. *Construction Management and Economics, 28*(4), 393–402.

Dzawu, M. M. (2012). Ghana's capital dries up as water system lags oil-led boom. *Bloomsberg Business*. 4th April. Available from: http://www.bloomberg.com/news/articles/2013-04-04/ghana-s-capital-dries-up-as-water-system-lags-oil. Accessed on 11 Jan 2015.

Efficiency Unit. (2001). *Serving the community by using the private sector*. Hong Kong: HKSAR Government.

Efficiency Unit. (2003). *Serving the community by using the private sector – An introductory guide to public private partnerships (PPPs)*. Hong Kong: Hong Kong Special Administrative Region Government.

Efficiency Unit. (2007). *Serving the community by using the private sector policy and practice* (2nd ed.). Hong Kong: The Hong Kong Special Administrative Region Government.

Efficiency Unit. (2012). *Serving the community by using the private sector – An introductory guide to Public Private Partnerships*. Hong Kong: The Hong Kong Special Administrative Region Government.

Eggers and Startup. (2006). *Closing the infrastructure gap: The role of PPPs*. Deloitte: Deloitte Research Study.

Ejohwomu, A. O., Oshodi, O. S., & Onifade, M. K. (2016). Causes of conflicts in construction projects in Nigeria: Consultants and contractors perspective. *Nigerian Journal of Technology, 35*(2), 270–277.

El-Gohary, N. M., Osman, H., & El-Diraby, T. E. (2006). Stakeholder management for public private partnerships. *International Journal of Project Management, 24*(7), 595–604. https://doi.org/10.1016/j.ijproman.2006.07.009.

English, L. M. (2006). Public private partnerships in Australia: An overview of their nature, purpose, incidence and oversight. *University of New South Wales Law Journal, 29*, 250.

Esty, B. C. (2001). Structuring loan syndicates: A case study of the Hong Kong Disneyland project loan. *Journal of Applied Corporate Finance, 14*(3), 80–95.

European Centre for Development Policy Management (ECDPC). (2012). Africa at a turning point? The case of Mozambique. *GREAT Insights, 1*(10).

Ezulike, E. I., Perry, J. G., & Hawwash, K. (1997). The barriers to entry into the PFI Market. *Engineering, Construction and Architectural Management, 4*(3), 179–193.

Fall, M., Marin, P., Locussol, A., & Verspyck, R. (2009). *Reforming urban water utilities in Western and Central Africa: Experiences with public-private partnerships – Volume 1: Impact and lessons learned* (Discussion Paper Series 13). Washington, DC: World Bank.

Fang, D. P., Xie, F., Huang, X. Y., & Li, H. (2004). Factor analysis-based studies on construction workplace safety management in China. *International Journal of Project Management, 22*(1), 43–49.

Farlam, P. (2005). *Working together. Assessing public-private partnerships in Africa, Nepad Policy* (Focus Series). Pretoria: The South African Institute of International Affairs (SAIIA).

Farquharson, E., de Mästle, C. T., & Yescombe, E. R. (2011). *How to engage with the private sector in public-private partnerships in emerging markets*. Washington, DC: World Bank Publications.

Faure, A. (2015). *Public international law debate concerning forced evictions in China*. Browser Download This Paper.

Field, A. P. (2013). Discovering statistics using IBM SPSS Statistics. London: Sage publication.

Field, J. E., & Peck, E. (2003). *Public–private partnerships in healthcare: The managers' perspective. Health & Social Care in the Community, 11*(6), 494–501.

Fox, P., & Skitmore, M. (2007). Factors facilitating construction industry development. *Building Research and Information, 35*(2), 178–188.

Frilet, M. (1997). Some Universal Issues in BOT Projects for Public Infrastructure. *The International Construction Law Review, 14*(4), 499–512.

Fuest, V., & Haffner, S. A. (2007). PPP-policies, practices and problems in Ghana's urban water supply. *Water Policy, 9*(2), 169–192.

Gannon, M. J., & Smith, N. J. (2011). An effective outline business case to facilitate successful decision-making. *Construction Management and Economics, 29*(2), 185–197.

Gardiner, P. D., & Simmons, J. E. (1992). Analysis of Conflict and Change in Construction Projects. *Construction Management and Economics, 10*(6), 459–478.

Gboyega, A. (2013). Lekki-Epe Road: Lagos to raise N87.5bn to acquire concession rights. *This Day Live,* 29 August. Available from: http://www.thisdaylive.com/articles/lekki-epe-road-lagos-to-raise-n87-5bn-to-acquire-concession-rights/157543/. Accessed on 2 Jan 2015.

German International Cooperation (GIZ). (2013). *Cooperation with the private sector in Ghana: Country report 2013*. Available from: http://www.giz.de/expertise/downloads/giz2013-en-ghana-country-report.pdf

Ghana News Agency. (2010). *Issues criticized in STX housing deal not addressed – Minority*. Available from: http://www.ghananewsagency.org/print/18673. Accessed on 5 Nov 2014.

Ghana News Agency (GNA). (2012). *ILGS/GIZ awards excelling MMDAs*. Available from: ghananewsagency.org/politics/ilgs-giz-awards-excelling-mmdas-38725. Accessed on 5 Oct 2014.

Gifford, J. (2013). *Public private partnerships: Prospect for the future international transport forum, Leipzig, Germany*. Available from: http://www.ppptransport.eu/docs/ITF2013/3_Gifford.pdf. Accessed May 2015.

Gorse, C. A. (2003). Conflict and conflict management in construction. In *Proceedings of 19th annual conference of the Association of Researchers in Construction Management* (pp. 173–182).

Graphic Online. (2011). *Projects & developments: Kumasi Central Market Project in Limbo*. Available from: http://www.modernghana.com/news/318409/1/kumasi-central-market-project-in-limbo.html. Accessed on 10 Jan 2015.

Griffs, M. (2011). *Economic indicators for dummies*. Hoboken: Wiley.

Grimsey, D., & Lewis, M. K. (2004). *Public and private partnerships. The worldwide revolution in infrastructure provision and project finance* (p. 01060-3815). Northampton: Edward Elgar Publishers.

Grimsey, D., & Lewis, M. (2007). *Public private partnerships and public procurement. Agenda, 14*(2), 171–188.

Gupta, U. G. (1991). *Validating and verifying knowledge-based systems*. New York: Institute of Electrical and Electronics Engineers (IEEE) Computer Society.

Gyasiwaa, A. (2011). Akyena Brentuo: STX deal will happen. *Myjoyonline*, August 22. Available from: http://politics.myjoyonline.com/pages/news/201108/71540.php. Accessed on 4 Nov 2014.

Harmon, K. M. (2003). Conflicts between owner and contractors: proposed intervention process. *Journal of Management in Engineering, 19*(3), 121–125.

Hayllar, M. R. (2010). Public-private partnerships in Hong Kong: Good governance–The essential missing ingredient? *Australian Journal of Public Administration, 69*(1), 99–119.

HM Treasury and Infrastructure U.K. (2013). *PFI projects data: March 2013*. Available from: https://www.gov.uk/government/statistics/pfi-projects-data-march-2013. Aug 2014.

Ho, P. H. (2006). Development of public private partnerships (PPPs) in China. *International Roundup, 19*(1).

Hodge, J. (2003). *Unsolicited proposals: The issues of private infrastructure projects*. Washington, DC: World Bank Publications. http://siteresources.worldbank.org/EXTFINANCIALSECTOR/Resources/282884-1303327122200/257Hodge-031103.pdf. Mar 2017.

Hodge, G. A., & Greve, C. (2005). *The challenge of public-private partnerships: Learning from international experience*. U.K: Edward Elgar Publishing.

Hodge, G. A., Greve, C., & Boardman, A. E. (Eds.). (2010). *International handbook on public-private partnership*. Northampton: Edward Elgar Publishing.

Hodges, J., & Dellacha, G. (2007). *Unsolicited infrastructure proposals: How some countries introduce competition and transparency PPIAF working paper no. 1, 2007*. http://www-wds.worldbank.org/external/default/WDSContentServer/WDSP/IB/2007/12/06/000020439_20071206143329/Rendered/PDF/417300Unsolici1als0PPIAF0101 PUBLIC1.pdf. Mar. 2017.

Hong Kong Disneyland Resort Media Relations (HKDRMR). (2014). *Walt Disney parks and resorts*. Available from http://hkcorporate.hongkongdisneyland.com/hkdlcorp/en_US/home/home?name=HomePage. Accessed Feb 2015.

Hsu, T. H., & Yang, T. S. (1997). The application of fuzzy synthetic decision to the human resource management. *Fu Jen Management Review, 4*(2), 85–100.

Hu, Y., Chan, A. P., Le, Y., Xu, Y., & Shan, M. (2016). Developing a program organization performance index for delivering construction megaprojects in China: Fuzzy synthetic evaluation analysis. *Journal of Management in Engineering, 32*, 05016007.

Huang, S., & Hsu, C. H. C. (2005). Mainland Chinese residents' perceptions and motivations of visiting Hong Kong: Evidence from focus group interviews. *Asia Pacific Journal of Tourism Research, 10*(2), 191–205.

Hudson, L. A., & Ozanne, J. L. (1988). Alternative ways of seeking knowledge in consumer research. *Journal of Consumer Research, 14*, 508–521.

Hurwitz, L. (1973). Comparative politics. *Special Issue on Revolution and Social Change, 5*(3), 449–463.

Hwang, B. G., Zhao, X., & Gay, M. J. S. (2013). Public private partnership projects in Singapore: Factors, critical risks and preferred risk allocation from the perspective of contractors. *International Journal of Project Management, 31*(3), 424–433.

Hwang, B. G., Zhu, L., & Tan, J. S. H. (2017). Identifying critical success factors for green business parks: case study of Singapore. *Journal of Management in Engineering, 33*(5). http://ascelibrary.org/doi/full/10.1061/%28ASCE%29ME.1943-5479.0000536.

Ibem, E. O. (2011). The contribution of Public–Private Partnerships (PPPs) to improving accessibility of low-income earners to housing in southern Nigeria. *Journal of Housing and the Built Environment, 26*(2), 201–217.

Ibrahim, A. D., Price, A. D. F., & Dainty, A. R. J. (2006). The analysis and allocation of risks in public-private partnerships in infrastructure projects in Nigeria. *Journal of Financial Management of Property and Construction, 11*(3), 149–163.

Iheduru, O. M. (Ed.). (2001). *Contending issues in African development: Advances, challenges, and the future (No. 219)*. Westport: Greenwood Publishing Group.

Infrastructure Concession Regulatory Commission (ICRC). (2012). *PPP project case studies*. Available from: http://ppptoolkit.icrc.gov.ng/ppp-project-case-studies/. Accessed on 18 Jan 2015.

International Finance Corporation (IFC). 2014. *Asutsuare water treatment plant, summary of investment information* [cited 2014 Oct 7]. Available from: http://www.ifc.org/ifcext/spiwebsite1.nsf/651aeb16abd09c1f8525797d006976ba/9062cbcf30cf532385257b3b0065cc41?

International Monetary Fund. (2010a). *Cameroon: Report on the observance of standards and codes— Fiscal transparency module*. Country Report No. 10/264, Washington, DC.

International Monetary Fund. (2010b). *Regional economic outlook: Sub-Saharan Africa resilience and risks*. Washington, DC: IMF.

Ismail, S. (2013a). Critical success factors of public private partnership (PPP) implementation in Malaysia. *Asia-Pacific Journal of Business Administration, 5*(1), 6–19.

Ismail, S. (2013b). Positive and negative factors for the successful adoption of Public Private Partnership: evidence from Malaysia. In *Public Private Partnerships (PPP): Malaysian studies* (pp. 65–81). Kuala Lumpur: IIUM Press.

Ismail, S., & Haris, A. F. (2014). Constraints in implementing public private partnership (PPP) in Malaysia. *Built Environment Project and Asset Management, 4*(3), 238–250.

Jacobson, C., & Choi, S. O. (2008). Success factors: public works and public-private partnerships. *International Journal of Public Sector Management, 21*(6), 637–657.

Jaffar, N., Tharim, A. A., & Shuib, M. N. (2011). Factors of conflict in construction industry: a literature review. *Procedia Engineering, 20*, 193–202.

Jamali, D. (2004). Success and failure mechanisms of public private partnerships (PPPs) in developing countries: Insights from the Lebanese context. *International Journal Public Sector Management, 17*(5), 414–430.

Javed, A. A. (2013). *A model of output specifications for public-private partnership projects* (Doctoral dissertation, The Hong Kong Polytechnic University).

Jefferies, M. (2006). Critical success factors of public private sector partnerships: A case study of the Sydney SuperDome. *Engineering, Construction and Architectural Management, 13*(5), 451–462.

Jefferies, M., Gameson, R., & Rowlinson, S. (2002). Critical success factors of the BOOT procurement system: Reflections from the stadium Australia case study. *Engineering Construction and Architectural Management, 9*(4), 352.

Jin, X. H., & Doloi, H. (2008). Interpreting risk allocation mechanism in public–private partnership projects: an empirical study in a transaction cost economics perspective. *Construction Management and Economics, 26*(7), 707–721.

Jooste, S. F. (2010). *Exploring the networks of organizations that enable and govern infrastructure public-private partnerships: An organizational field-level perspective*. Stanford: Stanford University.

Ke, Y., Wang, S., Chan, A. P., & Cheung, E. (2009). Research trend of public-private partnership in construction journals. *Journal of Construction Engineering and Management, 135*(10), 1076–1086.

Ke, Y., Wang, S., & Chan, A. P. C. (2010a). Risk allocation in public-private partnership infrastructure projects: Comparative study. *Journal of Infrastructure Systems, 16*, 343.

Ke, Y., Wang, S., Chan, A. P., & Lam, P. T. (2010b). Preferred risk allocation in China's public–private partnership (PPP) projects. *International Journal of Project Management, 28*(5), 482–492.

Ke, Y., Wang, S., Chan, A. P., & Cheung, E. (2011). Understanding the risks in China's PPP projects: ranking of their probability and consequence. *Engineering, Construction and Architectural Management, 18*(5), 481–496.

Khosravi, S., & Afshari, H. (2011). *A success measurement model for construction projects.* International conference on financial management and economics. IPRD, Vol. 11. Singapore: IACSIT Press.

Konchar, M., & Sanvido, V. (1998). Comparison of U.S. project delivery systems. *Journal of Construction Engineering and Management, 124*(6), 435–444.

Kumaraswamy, M. M. (1997). Conflicts, claims and disputes in construction. *Engineering, Construction and Architectural Management, 4*(2), 95–111.

Kumaraswamy, M. M., & Morris, D. A. (2002). Build-operate-transfer-type procurement in Asian megaprojects. *Journal of Construction Engineering and Management, 128*(2), 93–102.

Kumaraswamy, M. M., & Zhang, X. Q. (2001). Governmental role in BOT-led infrastructure development. *International Journal of Project Management, 19*(4), 195–205.

Kwak, Y. H., Chih, Y., & Ibbs, C. W. (2009). Towards a comprehensive understanding of public private partnerships for infrastructure development. *California Management Review, 51*(2), 51–78.

Lagos State PPP Office. (2010). *Lagos Infrastructure Project (LIP) Concession.* Available from: http://www.lagosstateppp.gov.ng/projects/project_portfolio/project_in_construction.asp. Accessed on 20 Jan 2015.

Lai, I. K. W., & Lam, F. K. S. (2010). Perception of various performance criteria by stakeholders in the construction sector in Hong Kong. *Construction Management and Economics, 28*(4), 377–391.

Lam, P., & Javed, A. (2015). Comparative study on the use of output specifications for Australian and U.K. PPP/PFI Projects. *Journal of Performance of Constructed Facilities.* https://doi.org/10.1061/(ASCE)CF.1943-5509.0000554.

Lam, E. W., Chan, A. P., & Chan, D. W. (2008). Determinants of successful design-build projects. *Journal of Construction Engineering and Management, 134*(5), 333–341.

Larbi C. (2014). Solving Ghana's infrastructure deficit…The role Of PPP. *Daily Guide.* 18 October. Available from: www.dailyguideghana.com/solving-ghanas-infrastructure-deficit-the-role-of-ppp. Accessed on 9 Oct 2014.

Laryea, S., & Hughes, W. (2008). How contractors price risk in bids: theory and practice. *Construction Management and Economics, 26*(9), 911–924.

Lee, V. (2005). *Public – Private Partnerships.* Hong Kong: *Research and Library Services Division Legislative Council Secretariat.*

Legislative Council. (1999). *Legislative Council Brief, Road Tunnels (Government) Ordinance (Cap. 368).* Hong Kong Government. Accessed from: http://www.legco.gov.hk/yr98-99/english/bc/bill_82/general/82_brf.pdf. Accessed on 3 Dec 2014.

Legislative Council. (2009). *Panel on information technology and broadcasting meeting on 9 March 2009 Updated background brief on Cyberport Project, Hong Kong Government.* Accessed from: http://www.legco.gov.hk/yr08-09/english/panels/itb/papers/itb0309cb1-955-4-e.pdf

Legislative Council Secretariat. (2006). *The adoption of the build-operate-transfer model and its variants in developing transport infrastructures in selected places research and library services division.* http://www.legco.gov.hk/yr05-06/english/sec/library/0506in13e.pdf. Accessed on 10 Jan 2015.

Leiringer, R. (2003). *Technological innovations in the context of public-private partnership projects.* Unpublished PhD thesis, Royal Institute of Technology, Stockholm.

Lekki Concession Company (LCC). (2015). *Project scope.* Available from: http://www.lcc.com.ng/tolls.asp?pid=21. Accessed on 18 Jan 2015.

Leung, B. Y., & Hui, E. C. (2002). *Option pricing for real estate development: Hong Kong Disneyland. Journal of Property Investment & Finance, 20*(6), 473–495.

Levy, S. M. (1996). *Build, operate, transfer: Paving the way for tomorrow's infrastructure.* New York: Wiley.

Li, B. (2003). *Risk management of construction public private partnership projects*. Ph.D. thesis, Glasgow Caledonian University, UK.

Li, B., Akintoye, A., Edwards, P. J., & Hardcastle, C. (2005a). Critical success factors for PPP/PFI projects in the UK construction industry. *Construction Management and Economics, 23*(5), 459–471.

Li, B., Akintoye, A., Edwards, P. J., & Hardcastle, C. (2005b). Perceptions of positive and negative factors influencing the attractiveness of PPP/PFI procurement for construction projects in the UK: Findings from a questionnaire survey. *Engineering Construction Architectural Management, 12*(125), 148.

Li, B., Akintoye, A., Edwards, P. J., & Hardcastle, C. (2005c). The allocation of risk in PPP/PFI construction projects in the UK. *International Journal of project management, 23*(1), 25–35.

Lim, C. S., & Mohamed, M. Z. (1999). Criteria of project success: an exploratory re- examination. *International Journal of Project Management, 17*, 243–248.

Ling, F. Y. Y., Chan, S. L., Chong, E., & E, L. P. (2004). Predicting performance of design-build and design-bid-build projects. *Journal of Construction Engineering and Management, 130*(1), 75–83.

Lingard, H., & Rowlinson, S. (2006). Letter to the editor. *Construction Management Economics, 24*(1107), 1109.

Liu, T., & Wilkinson, S. (2011). Adopting innovative procurement techniques: Obstacles and drivers for adopting public private partnerships in New Zealand. *Construction Innovation: Information, Process, Management, 11*(4), 452–469.

Liu, T., & Wilkinson, S. (2013). Can the pilot public-private partnerships project be applied in future urban rail development? A case study of Beijing Metro Line 4 project. *Built Environment Project and Asset Management, 3*(2), 250–263.

Liu, J., Li, Q., & Wang, Y. (2013). Risk analysis in ultra-deep scientific drilling project—A fuzzy synthetic evaluation approach. *International Journal of Project Management, 31*, 449–458.

Liu, J., Love, P. E., Smith, J., Regan, M., & Davis, P. R. (2014). Life cycle critical success factors for public-private partnership infrastructure projects. *Journal of Management in Engineering, 31*, 04014073.

Liu, J., Love, P., Davis, P., Smith, J., & Regan, M. (2015). Conceptual framework for the performance measurement of public-private partnerships. *Journal of Infrastructure Systems*. https://doi.org/10.1061/(ASCE)IS.1943-555X.0000210.

Liu, T., Wang, Y., & Wilkinson, S. (2016). Identifying critical factors affecting the effectiveness and efficiency of tendering processes in Public–Private Partnerships (PPPs): A comparative analysis of Australia and China. *International Journal of Project Management, 34*(4), 701–716.

Liyanage, C., & Villalba-Romero, F. (2015). Measuring success of ppp transport projects: a cross-case analysis of toll roads. *Transport Reviews, 35*(2), 140–161.

Lo, T. Y., Fung, I. W., & Tung, K. C. (2006). Construction delays in Hong Kong civil engineering projects. *Journal of Construction Engineering and Management, 132*(6), 636–649.

Loosemore, M., Hall, K., & Dainty, A. (1996). *Innovation and courage in construction management research*, In: Thorpe, A (Ed), 12th annual, -IRCOM conference.

Loxley, J. (2013). Are public–private partnerships (PPPs) the answer to Africa's infrastructure needs? *Review of African Political Economy, 40*(137), 485–495.

Lucko, G., & Rojas, E. (2010). Research validation: Challenges and opportunities in the construction domain. *Journal of Construction Engineering and Management*, 127–135. https://doi.org/10.1061/(ASCE)CO.1943-7862.0000025.

Macauhub. (2006). *Businesspeople protest against container inspection taxes at Maputo port*. Retrieved January 19, 2014, from http://www.macauhub.com.mo/en/2006/12/26/2262/

Madura, J. (2012). *Financial markets and institutions*. Boston: Cengage Learning.

Mahato, K. B., & Ogunlana, S. O. (2011). Conflict dynamics in a dam construction project: A case study. *Built Environment Project and Asset Management, 1*(2), 176–194.

Mainelli, M. (2003). PFI and PPP: Could they result In Enron UK? *Balance Sheet, 11*(2).

Mak, T. P. L. (2012). *The application of private public partnership in Hong Kong*. HKU Theses Online (HKUTO).

Mak, C. K., & Mo, S. (2005). *Some aspects of the PPP approach to transport infrastructure development in Hong Kong*. Highways Department, Government of the Hong Kong Special Administrative Region.

Marin, P. (2009). *Public-private partnerships for urban water utilities: A review of experiences in developing countries*. Washington, DC: World Bank Publications.

Marques, R. C. (2018). Is arbitration the right way to settle conflicts in PPP arrangements? *Journal of Management in Engineering, 34*(1), 05017007.

Matusitz, J. (2011). Disney's successful adaptation in Hong Kong: A glocalization perspective. *Asia Pacific Journal of Management, 28*, 667–681.

Memorandum of Understanding (MOU). (2010). Available from: http://danquahinstitute.org/docs/STX-MOU.pdf. Accessed on 17 Nov 2014.

Meng, X., Zhao, Q., & Shen, Q. (2011). Critical success factors for transfer-operate-transfer urban water supply projects in China. *Journal of Management in Engineering*. https://doi.org/10.1061/(ASCE)ME.1943-5479.0000058.

Metro Planning Coordinating Unit (MPCU). (2013). *Sekondi Takoradi Metropolitan Assembly Annual Progress Report 2012*. Available from: www.ndpc.gov.gh/.../Dist%20and%20Sec%20APR%202012/DISTRICT. Accessed on 20 Sept 2014.

Middleton, N. (2000). Public private partnerships — a natural successor to privatisations? http://www.pwcglobal.com/uk/eng/about/svcs/pfp/ppp.html. Accessed on Oct 2013.

Mingers, J., & Brocklesby, J. (1996). Multimethodology: towards a framework for critical pluralism. *Systemist, 18*(3), 101–132.

Mingle J. (2010). *Commentary on STX housing project*. Ghana Broadcasting Corporation. Available from: http://gbcghana.com/kitnes/cache/dynamic/1156360.htm. Accessed on 25 Oct 2014.

Ministry of Finance and Economic Planning. (2011). *National policy on public – private partnerships*. Ghana.

Ministry of Finance and Economic Planning. (2014a). *Pipeline projects: Ghana public private partnerships programme*. Government of Ghana, Ghana. Available from: www.mofep.gov.gh/sites/.../Pipeline-Projects-Ghana-PPP-11-08-14.pdf. Accessed on 25 Sept 2014.

Ministry of Finance and Economic Planning. (2014b). *Ghana public private partnerships programme: Newsletter*. Government of Ghana, Ghana. Available from: www.mofep.gov.gh/sites/default/files/docs/.../PPP-Newsletter-042014.pdf. Accessed on 25 Sept 2014.

Mitkus, S., & Mitkus, T. (2014). Causes of conflicts in a construction industry: A communicational approach. *Procedia-Social and Behavioral Sciences, 110*, 777–786.

Mladenovic, G., Vajdic, N., Wundsch, B., & Temeljotov-Salaj, A. (2013). Use of key performance indicators for PPP transport projects to meet stakeholders' performance objectives. *Built Environment Project and Asset Management, 3*(2), 228–249.

Mohr, J., & Spekman, R. (1994). Characteristics of partnership success: partnership attributes, communication behaviour, and conflict resolution techniques. *Strategic Management Journal, 15*, 135–152.

Molenaar, K. R., & Songer, A. D. (1998). Model for public sector design-build project selection. *Journal of Construction Engineering and Management, 124*(6), 467–479.

Monsalve, C. (2009). *Private participation in transport: Lessons from recent experience in Europe and Central Asia*. Washington, DC: PPIAF, World Bank.

Morledge, R., & Owen, K. (1998). *Critical success factors in PFI projects*. Proceedings of the 14th ARCOM Annual Conference, (AAC' 98), University of Reading, pp. 565–574.

Morrison, M. R. (Ed.). (2012). *The principles of project finance*. Gower Publishing Ltd.

Mustafa, A. (1999). Public-private partnership: an alternative institutional model for implementing the private financial initiative in the provision of transport infrastructure. *The Journal of Project Finance, Summer*, 64–79.

Mustapha, A. R., & Whitfield, L. (2009). *Turning points in African democracy*. Rochester: Boydell & Brewer.

Nataraj, G. (2007), *Infrastructure challenges in South Asia: the role of public-private partnerships. No. 80*. ADB Institute Discussion Papers.

Neves, P., & Kim, D. J. (2017). *Managing unsolicited proposals in infrastructure: 5 key questions for governments*. World Bank Publication. http://blogs.worldbank.org/ppps/managing-unsolicited-proposals-infrastructure-5-key-questions-governments?CID=PPP_E_PPPNewsletter_EXTi

Ng, J. Y. K., & Chan, A. H. S. (2015). *The Ageing Construction Workforce in Hong Kong: A Review*. Proceedings of the International Multi-Conference of Engineers and Computer Scientists 2015 Vol II, IMECS 2015, March 18–20, 2015, Hong Kong.

Ng, S. T., Wong, Y. M., & Wong, J. M. (2012). Factors influencing the success of PPP at feasibility stage – A tripartite comparison study in Hong Kong. *Habitat International, 36*(4), 423–432.

Nisar, T. M. (2013). Implementation constraints in social enterprise and community Public Private Partnerships. *International Journal of Project Management, 31*(4), 638–651.

Norris, S., & Ogunbiyi, C. (2003). Letting the crown jewels fall into private hands: A case study of the Maputo port project. *The Journal of Structured Finance, 9*(2), 47–52.

Norusis, M. J. (2008). *SPSS 15.0 guide to data analysis*. Upper Saddle River, NJ: Prentice Hall.

Nwangwu, G. A. (2013). *A risk based approach to enhancing public-private partnership (PPP) projects in Nigeria*. PhD thesis, University of Hull, UK.

O'Karo T. J. (2011). *The Lekki Toll Gate Protest – An eyewitness account*. Bella Naija, 19 December. Available from: http://www.bellanaija.com/2011/12/19/the-lekki-toll-gate-protest-an-eyewitness-account/. Accessed on 10 Dec 2014.

Off-Taker Agreement. (2009). Available from: danquahinstitute.org/docs/STX-Off-TakerAgreement.pdf. Accessed on 12 Nov 2014.

Oforiokuma O. (2011). *PPPs and sub nationals – Lessons from around the World: Lekki Toll Road Concession, Lagos, Nigeria*. IFC Regional PPP Conference 14th November 2011. Available from: http://www.ifc.org/wps/wcm/connect/345f92004956d9af933bbf849537832d/9.b2_CaseStudiesTollRoads-LekkiConcession.pdf?MOD=AJPERES. Accessed on 15 Jan 2015.

Olufowose, E. (2008). *Infrastructure finance in Nigeria: Air and water transport*. 2nd annual banking and finance conference of the Chartered Institute of Bankers of Nigeria (CIBN) 28th–30th October, Abuja, pp. 1–3.

Oluwasanmi, O., & Ogidi, O. (2014). Public private partnership and Nigerian economic growth: Problems and prospects. *International Journal of Business and Social Science, 5*(11).

Organisation for Economic Co-operation and Development. (2008). *Public-private partnerships: In pursuit of risk sharing and value for money*. Paris: OECD Publications.

Organisation for Economic Co-operation and Development. (2010). *Dedicated public-private partnership units: A survey of institutional and governance structures*. Paris: OECD Publishing.

Osborne, S. (2002). *Public-private partnerships: Theory and practice in international perspective*. New York: Routledge.

Osei-Kyei, R., & Chan, A. P. C. (2015a). Developing transport infrastructure in Sub-Saharan Africa through public–private partnerships: Policy practice and implications. *Transport Reviews*, 1–17.

Osei-Kyei, R., & Chan, A. P. C. (2015b). Review of studies on the critical success factors for public–private partnership (PPP) projects from 1990 to 2013. *International Journal of Project Management, 33*(6), 1335–1346.

Osei-Kyei, R., & Chan, A. P. C. (2016). Implementing public–private partnership (PPP) policy for public construction projects in Ghana: critical success factors and policy implications. *International Journal of Construction Management*. https://doi.org/10.1080/15623599.2016.1207865.

Osei-Kyei, R., & Chan, A. P. (2017a). Factors attracting private sector investments in public–private partnerships in developing countries. *Journal of Financial Management of Property and Construction*.

Osei-Kyei, R., & Chan, A. P. (2017b). Empirical comparison of critical success factors for public-private partnerships in developing and developed countries. *Engineering, Construction and Architectural Management*.

Osei-Kyei, R., & Chan, A. P. (2017c). Risk assessment in public-private partnership infrastructure projects: empirical comparison between Ghana and Hong Kong. *Construction Innovation*, 204–223.

Osei-Kyei, R., & Chan, A. P. (2017d). Developing a project success index for public–private partnership projects in developing countries. *Journal of Infrastructure Systems, 23*(4), 04017028.

Osei-Kyei, R., & Chan, A. P. (2018a). A best practice framework for public-private partnership implementation for construction projects in developing countries. *Benchmarking: An International Journal.*

Osei-Kyei, R., & Chan, A. P. (2018b). Comparative study of governments' reasons/motivations for adopting public-private partnership policy in developing and developed economies/countries. *International Journal of Strategic Property Management, 22*(5), 403–414.

Osei-Kyei, R., Dansoh, A., & Ofori–Kuragu, J. K. (2014). Reasons for adopting Public–Private Partnership (PPP) for construction projects in Ghana. *International Journal of Construction Management, 14*(4), 227–238.

Osei-Kyei, R., Chan, A. P. C., Javed, A. A., & Ameyaw, E. E. (2016). Critical success criteria for public-private partnership projects: International experts' opinion. *International Journal of Strategic Property Management, 21.*

Osei–Kyei, R., Chan, A. P. C., Javed, A. A., & Ameyaw, E. E. (2017). Critical success criteria for public-private partnership projects: International experts' opinion. *International Journal of Strategic Property Management, 21.* https://doi.org/10.3846/1648715X.2016.1246388.

Osei-Kyei, R., Chan, A. P., Dansoh, A., Ofori-Kuragu, J. K., & Oppong, G. D. (2018a). Strategies for effective management of unsolicited public–private partnership proposals. *Journal of Management in Engineering, 34*(3), 04018006.

Osei-Kyei, R., Chan, A. P., Dansoh, A., Ofori-Kuragu, J. K., & Owusu, E. K. (2018b). Motivations for adopting unsolicited proposals for public-private partnership project implementation: a survey of international experts. *Journal of Financial Management of Property and Construction*, 221–238.

Osei-Kyei, R., Chan, A. P., Yu, Y., Chen, C., & Dansoh, A. (2019a). Root causes of conflict and conflict resolution mechanisms in public-private partnerships: Comparative study between Ghana and China. *Cities, 87*, 185–195.

Osei-Kyei, R., Chan, A. P., Yao, Y., & Mazher, K. M. (2019b). Conflict prevention measures for public-private partnerships in developing countries. *Journal of Financial Management of Property and Construction*, 39–57.

Owen, G., & Merna, A. (1997). The private finance initiative. *Engineering Construction and Architectural Management, 4*(3), 163–177.

Owusu W. Y. (2011). STX cheques bounced. *Daily Guide*, 22 August. Available from: http://www.modernghana.com/news/346347/1/stx-cheques-bounced.html. Accessed on 1 Nov 2014.

Oxford Business Group. (2011a). *The report: Ghana 2011.* London: Oxford Business Group Publications.

Oxford Business Group. (2011b). *Ghana: No place like home*. Available from: http://www.oxfordbusinessgroup.com/news/ghana-no-place-home. Accessed on 3 Oct 2014.

Oxford Business Group. (2012). *The report: Nigeria 2012.* London: Oxford Business Group Publications.

Oyedele, L. O. (2012). Avoiding performance failure payment deductions in PFI/PPP projects: model of critical success factors. *Journal of Performance of Constructed Facilities, 27*(3), 283–294.

Oyedele O. A. (2014). *Public-Private Partnership (PPP) and infrastructure provision in Nigeria.* AARC HES Conference.

Ozdoganm, I. D., & Birgonul, M. T. (2000). A decision support framework for project sponsors in the planning stage of build-operate-transfer (BOT) projects. *Construction Management and Economics, 18*(3), 343–353.

Pallister, J., & Law, J. (2006). *A dictionary of business and management.* Oxford: Oxford University Press.

Parfitt, M. K., & Sanvido, V. E. (1993). Checklist of critical success factors for building projects. *Journal of Management in Engineering, 9*(3), 243–249.

Pessoa, A. (2010). *Reviewing Public-private sector partnerships performance in developing countries* (No. 362, working paper). Universidade do Porto, Faculdade de Economia do Porto.

Plummer, J. (Ed.). (2013). *Focusing partnerships: A sourcebook for municipal capacity building in public-private partnerships*. Routledge.

Pollack, J. (2007). The changing paradigms of project management. *International Journal of Project Management, 25*, 266–274.

Pongsiri, N. (2002). Regulation and public private partnerships. *International Journal of Public Sector Management, 15*(6), 487–495.

Potts, K., & Ankrah, N. (2014). *Construction cost management: Learning from case studies*. London: Routledge.

PPIAF (Public-Private Infrastructure Advisory Facility). (2009). *"Unsolicited proposals." Module 5: Implementing and Monitoring*. Washington, DC: PPIAF. https://ppiaf.org/sites/ppiaf.org/files/documents/toolkits/highwaystoolkit/6/pdf-version/5-42.pdf.

PPIAF (Public-Private Infrastructure Advisory Facility). (2014). Unsolicited proposals – An exception to public initiation of infrastructure PPPs. Washington, DC: PPIAF. http://ppp.worldbank.org/public-private-partnership/sites/ppp.worldbank.org/files/documents/PPIAF_UnsolicitedProposals_EN.pdf. Mar 2017.

PPP Center Foundation. (2013). *PPPs as a way to invest*. Available from http://samorzadowefo-rum.pl/wp-content/uploads/2013/10/IRENA-HERBST.pdf. Accessed Dec 2014.

Price Waterhouse Coopers. (2012). *Public-Private Partnerships in Malaysia*. Available from: http://www.pwc.com/my/en/assets/services/ppp-projects-in-malaysia.pdf. Accessed May 2015.

Public-Private Infrastructure Advisory Facility (PPIAF). (2009). *Toolkit for Public-Private Partnerships in roads & highways: N4 Toll Road from South Africa to Mozambique*. Available from: http://www.ppiaf.org/sites/ppiaf.org/files/documents/toolkits/highwaystoolkit/6/pdf-version/safricamozambique.pdf

Qiao, L., Wang, S. Q., Tiong, R. L. K., & Chan, T. S. (2001). Framework for critical success factors of BOT projects in China. *Journal of Project Finance, 7*(1), 53–61.

Raftery, J., McGeorge, D., & Walters, M. (1997). Breaking up methodological monopolies: a multi-paradigm approach to construction management research. *Construction Management and Economics, 15*, 291–297.

Raisbeck, P., & Tang, L. C. (2013). Identifying design development factors in Australian PPP projects using an AHP framework. *Construction Management and Economics, 31*(1), 20–39.

Reinhardt, W. (2011). The role of private investment in meeting US transportation infrastructure needs. *Public Works Financing, 260*, 1–31.

Renesch, J. (Ed.). (2002). *Leadership in a new era: Visionary approaches to the biggest crises of our time*. New York: Cosimo Inc.

Robinson, H. S., & Scott, J. (2009). Service delivery and performance monitoring in PFI/PPP projects. *Construction Management and Economics, 27*(2), 181–197.

Rockart, J. F. (1982). The changing role of the information systems executive: A critical success factors perspective. *Sloan Management Review, 24*(1), 3–13.

Rosenau, M. D. (1984). *Project management for engineers, Lifetime Learning Pub*. Ann Arbor: University of Michigan.

Roumboutsos, A., & Anagnostopoulos, K. P. (2008). Public–private partnership projects in Greece: risk ranking and preferred risk allocation. *Construction Management and Economics, 26*(7), 751–763.

Runeson, G. (1997). The role of theory in construction management research: comment. *Construction Management and Economics, 15*, 299–302.

Rwelamila, P. D., Fewings, P., & Henjewele, C. (2015). Addressing the missing link in PPP projects: What constitutes the public? *Journal of Management in Engineering, 31*(5), 04014085.

Sachs, T., Tiong, R., & Qing Wang, S. (2007). Analysis of political risks and opportunities in public private partnerships (PPP) in China and selected Asian countries: Survey results. *Chinese Management Studies, 1*(2), 126–148.

Sadiq, R., & Rodriguez, M. J. (2004). Fuzzy synthetic evaluation of disinfection by-products—a risk-based indexing system. *Journal of Environmental Management, 73*(1), 1–13.

Sahn, D. E., & Younger, S. (2009). *Living standards in Africa: Poverty. Initiative for policy dialogue working paper series*. Retrieved from http://policydialogue.org/files/publications/Ch_15_Sundaram_Tendulkar.pdf

Sandelowski, M. (1995). Qualitative analysis: What it is and how to begin. *Research in Nursing & Health, 18*(4), 371–375.
Sanni, A. O., & Hashim, M. (2014). *Building infrastructure through Public Private Partnerships in sub-Saharan Africa: Lessons from South Africa. Procedia-Social and Behavioral Sciences, 143*, 133–138.
Securities Africa. (2012). *Weekly African footprints*. Available from; http://www.securitiesafrica.com/Documents/Docs/Weekly_African_Footprint_13_July_2012.pdf. Aug 2014.
Sekondi – Takoradi Metropolitan Assembly City-Wide Settlement Upgrading Fund (STMA – CSUF). (2012). *Kojokrom market sheds project*. Available from: www.stma-csuf.org/pages/index.php?siteid=csuf&page=13029. Accessed on 8 Sept 2014.
Sengupta, U. (2006). Government intervention and public-private partnerships in housing delivery in Kolkata. *Habitat International, 30*(3), 448–461.
Seymour, D., & Rooke, J. (1995). The culture of the industry and the culture of research. *Construction Management and Economics, 13*, 511–523.
Shimron, B. (1979). Polybius on Rome: a re-examination of the evidence. *SCI5 (1979–80)*, 94–117.
Shen, L. Y., Platten, A., & Deng, X. P. (2006). Role of public private partnerships to manage risks in public sector projects in Hong Kong. *International Journal of Project Management, 24*(7), 587–594.
Shendy, R. (2013a). *Ghana- Ghana- PPP Project: P125595- Implementation Status Results Report Sequence 03*. Washington, DC: World Bank.
Shendy, R. (2013b). *Ghana – Ghana – PPP Project: P125595 – Implementation status results report: Sequence 04*. Washington, DC: World Bank.
Shendy, R. (2014). *Ghana – Ghana – PPP Project: P125595 – Implementation status results report: Sequence 05*. Washington, DC: World Bank.
Shendy, R., Kaplan, Z., & Mousley, P. (2011). *Toward better infrastructure: Conditions, constraints and opportunities in financing PPP in select African countries*. Washington, DC: World Bank Publications.
Simon, T. W., Feng, C., Nelson, L. P. (2016). *China's changing legal system Lawyers & Judges on Civil & Criminal Law*. US: Palgrave Macmillan.
Smyth, H. J., & Morris, P. W. (2007). An epistemological evaluation of research into projects and their management: methodological issues. *International Journal of Project Management, 25*(4), 423–436.
Soomro, M. A., & Zhang, X. (2013). Roles of private-sector partners in transportation public-private partnership failures. *Journal of Management in Engineering, 31*(4), 04014056.
Southern African Development Community (SADC). (2012). *SADC PPP Case studies: A case study on N4 Toll Concessio*n. Available from: http://www.thetradebeat.com/sadc-business-case-studies/n4-toll-Concession. Accessed on 15 Jan 2015.
Special Unit for South-South Cooperation (SUSSC). (2012). *South Africa and Mozambique – The N4 Toll road: Case study (Transportation)*. Available from: http://www.esc-pau.fr/ppp/documents/featured_projects/south_africa_and_mozambique.pdf
State Government of Victoria. (2015). *Public-Private Partnership. State of Victoria, Australia*. Available from: http://www.dtf.vic.gov.au/Infrastructure-Delivery/Public-private-partnerships. Accessed Jan 2015.
Statesman. (2011). *Bagbin: Yes, we have secured STX funds, from Stanchart, no, I mean, Barclays*. Available from: http://www.ghanaweb.com/GhanaHomePage/NewsArchive/artikel.php?ID=213071
Stein, S. W. (1994). Build-Operate-Transfer (BOT): A re-evaluation. *The International Construction Law Review, 11*(2), 101–113.
Stevens, J. P. (2012). *Applied multivariate statistics for the social sciences*. New York: Routledge.
Stromback L. (1993), *Baltzar von Platen, Thomas Telfordoch Göta Kanal: entreprenörskap och tekniköverföring i brytningstid*. Doctoral thesis, Symposion Graduale, Stehag, Sweden.
Szolnoki, G., & Hoffmann, D. (2013). Online, face-to-face and telephone surveys—Comparing different sampling methods in wine consumer research. *Wine Economics and Policy, 2*(2), 57–66.
Tam, C. M. (1999). Build-operate-transfer model for infrastructure developments in Asia: Reasons for successes and failures. *International Journal of Project Management, 17*(6), 377–382.

Tang, L., & Shen, Q. (2013). Factors affecting effectiveness and efficiency of analyzing stakeholders' needs at the briefing stage of public private partnership projects. *International Journal of Project Management, 31*(4), 513–521.

Tang, L., Shen, Q., Skitmore, M., & Cheng, E. W. (2012). Ranked critical factors in PPP briefings. *Journal of Management in Engineering, 29*(2), 164–171.

Taylor I. (2000). *Public-Private Partnerships: Lessons from the MDC Toll Road. Working Paper No. 00/44*. Development Policy Research Unit, University of Cape Town, SA.

Thomas, H. (2009). The local press and urban renewal: A South Wales case study. *International Journal of Urban and Regional Research, 18*(2), 315–333.

Thomas, A. V., Kalidindi, S. N., & Ananthanaraynan, K. (2003). Risk perception analysis of BOT road project participants in India. *Construction Management and Economics, 21*(4), 393–407.

Thomas, A. V., Kalidindi, S., & Ganesh, L. S. (2006). Modelling and assessment of critical risks in BOT road projects. *Construction Management and Economics, 24*, 407–424.

Tiong, R. L. (1996). CSFs in competitive tendering and negotiation model for BOT projects. *Journal of Construction Engineering and Management, 122*(3), 205–211.

Tiong, R. L. K., Yeo, K. T., & McCarthy, S. C. (1992). Critical Success Factors in Winning BOT Contracts. *ASCE Journal of Construction Engineering and Management, 118*(2), 217–228.

Toor, R. S., & Ogunlana, S. O. (2010). Beyond the 'iron triangle': Stakeholder perception of key performance indicators (KPIs) for large-scale public sector development projects. *International Journal of Project Management, 28*(3), 228–236.

Trading Economics. (2016a). *Hong Kong Inflation rate: Data from 1981–2016*. Available from: http://www.tradingeconomics.com/hong-kong/inflation-cpi. Accessed on 29 July 2016.

Trading Economics. (2016b). *Ghana Interest rate: Data from 2002–2016*. Available from: http://www.tradingeconomics.com/ghana/interest-rate. Accessed on 29 July 2016.

Trading Economics. (2016c). *Hong Kong Interest rate: Data from 1998–2016*. Available from: http://www.tradingeconomics.com/hong-kong/interest-rate. Accessed on 29 July 2016.

Trebilcock, M., & Rosenstock, M. (2013). Infrastructure PPPs in developing world: Lessons from recent experience mimeo.

Trinity International L.L.P. (2009). *A review of the Lekki-Epe Expressway Toll Road Concession Project*. Available from: http://www.trinityllp.com/a-review-of-the-lekki-epe-expressway-toll-road-concession-project/. Accessed on 21 Dec 2014.

Trinity International L.L.P. (2010). *Are we there yet? Sub-Saharan Africa transport projects*. Retrieved January 22, 2014, from http://www.trinityllp.com/are-we-there-yet-sub-saharan-africa-transport-projects

United Kingdom Essays. (2013). *The Procurement System Construction Essay* [online]. Available from: http://www.ukessays.com/essays/construction/the-procurement-system-construction-essay.php?cref=1. Accessed Aug 2014.

United Nations Development Programme (UNDP). (2017). *Implementation – Managing conflict: What are conflicts*, UNDP. Available from: http://pppue.undp.2margraf.com/en/20_3.htm. Accessed on Feb 2017.

United Nations Economic Commission for Europe. (2008). *Guide book on promoting good governance on public private partnership, UN, Geneva*. http://www.unece.org/ceci/publications/ppp.pdf. Accessed on 20 Dec 2013.

United Nations Human Settlements Programme. (2011). *Slum-upgrading facility – End of programme evaluation*. Evaluation Report 4/2011. Available from: www.unhabitat.org/evaluations. Accessed on 22 Sept 2014.

United Nations Human Settlements Programme (UN – Habitat). (2009). *The UN-HABITAT slum upgrading facility. Newsletter,* Issue No. 7. Available from: www.mirror.unhabitat.org/pmss/getElectronicVersion.aspx?nr=2639&alt=1. Accessed on 4 Sept 2014.

United State Agency International Development (USAID). (2005). *Building public private partnership: South Africa final report. USAID*. Available from: http://pdf.usaid.gov/pdf_docs/Pdacg387.pdf

United States Agency for International Development (USAID). (2008). *The South African PPP program: The Inkosi Albert Luthuli Hospital Transaction*. Available from: http://www.energytoolbox.org/library/ppp_in_infrastructure/presentations/The_South_Africa_PPP_Program.pdf. Accessed on Feb 2015.

United States Department of Transportation (US DOT). (2004). *Report to Congress on Public Private Partnership, U.S. DOT.* Available from: www.fhwa.dot.gov/reports/pppdec2004

Walker, C., & Smith, A. J. (1995). *Privatized infrastructure: The BOT approach*. London: Thomas Telford.

Wall, A., & Wall, T. (2013). *Public-private Partnerships in the USA: Lessons to be Learned for the United Kingdom*. New York: Routledge.

Wang, S. Q., Ke, Y. J., & Xie, J. (2012). *Public private partnership implementation in China*, in G.M.

Westerveld, E. (2003). The Project Excellence Model: linking success criteria and critical success factors. *International Journal of Project Management, 21*(6), 411–418.

Wibowo, A., & Mohamed, S. (2010). Risk criticality and allocation in privatised water supply projects in Indonesia. *International Journal of Project Management, 28*(5), 504–513.

Winch, G. M. (2000). Innovativeness in British and French construction: the evidence from Trans manche–Link. *Construction Management and Economics, 18*, 807–817.

Winch, G. M., Onishi, M., & Schmidt, S. (Eds.). *Taking stock of PPP and PFI around the World* (pp. 29–36). The Association of Chartered Certified Accountants (ACCA).

Wits Business School. (2007a). *The Public Private Partnership between the KwaZulu Natal Department of Health and Impilo Consortium in Inkosi Albert Luthuli Central Hospital.* Available from: http://www.ppp.gov.za/Legal%20Aspects/Case%20Studies/IALCH%20 final%20document%20comparison.pdf. Accessed Feb 2015.

Wits Business School. (2007b). *Case studies on the PPPs at Humansdorp District Hospital Universitas and Pelonomi Hospitals and Inkosi Albert Luthuli Central Hospital.* Available from:http://www.ppp.gov.za/Legal%20Aspects/Case%20Studies/Humansdorp%20 Overall%20findings.pdf. Accessed Feb 2015.

Wong, A. (2005). ICAC to take role in public private deal. *The Standard*, October 31. http://www.thestandard.com.hk/news_print.asp?art_id=4529&sid=5255280. Accessed on May 2015.

World Bank. (2009). *Attracting investors to African public private partnerships: A project preparation guide*. Washington, DC: World Bank Publications.

World Bank. (2011). *PPIAF assistance in Ghana. Public-Private Infrastructure Advisory Facility (PPIAF)*. Washington, DC: The World Bank.

World Bank. (2012). *Ghana – Public Private Partnership Project (PPP)*. Washington, DC: World Bank.

World Bank. (2013a). *Infrastructure Policy Unit 2012 Latin America and the Caribbean PPI Data Update, (Note 93)*. Washington, DC: World Bank Publications.

World Bank. (2013b). *Infrastructure Policy Unit 2012 East Asia and the Pacific PPI Data Update. (Note 91)*. Washington, DC: World Bank publications.

World Bank. (2013c). *Infrastructure Policy Unit 2012 Africa PPI Data Update. (Note 90)*. Washington, DC: World Bank publications.

World Bank. (2014a). *2014 Global PPI Updates*. Washington, DC: World Bank Publications.

World Bank. (2014b). *Building integrated markets within the East African Community: EAC opportunities in Public-Private Partnership approaches to the region's infrastructure needs*. Washington, DC: World Bank Publications.

World Bank (2014c). *Ghana private participation in infrastructure projects database; country snapshots* [cited 2014 Oct 11]. Available from: http://ppi.worldbank.org/snapshots/country/ghana

World Bank. (2014d). *Unsolicited proposals – An exception to public initiation of infrastructure PPPs: An analysis of global trends and lesson learned*. Washington, DC: World Bank Publication.

World Bank. (2014e). *Overcoming constraints to the financing of infrastructure: Success stories and lessons learned: Country, sector and project examples of overcoming constraints to the financing of infrastructure*. Washington, DC: World Bank Group.

World Bank. (2015a). *Private Participation in Infrastructure (PPI), regional snapshots*. Washington, DC: World Bank. Available from: http://ppi.worldbank.org/explore/ppi_exploreRegion.aspx?regionID=1. Accessed on May 2015.

World Bank. (2015b). *Private Participation in Infrastructure (PPI), Middle East and North Africa. Regional snapshots.* Washington, DC: World Bank. Available from: http://ppi.worldbank.org/explore/ppi_exploreRegion.aspx?regionID=1. Accessed on May 2015.

World Bank. (2015c). *Data: Population total.* Available from: http://data.worldbank.org/country

World Bank. (2017). *Guidelines for the development of a policy for man-aging unsolicited proposals in infrastructure projects.* Washington, DC. http://pubdocs.worldbank.org/en/976121488983070966/Guidelines-for-the-Development-of-a-USP-Policy.pdf. May 2017.

World Economic Forum. (2010). *Paving the way: Maximizing the value of private finance in infrastructure.* World Economic Forum, USA. Available from: http://www3.weforum.org/docs/WEF_IV_PavingTheWay_Report_2010.pdf

Xenidis, Y., & Angelides, D. (2005). The financial risks in build-operate-transfer projects. *Construction Management and Economics, 23*, 431–441.

Xiao, H., & Proverbs, D. (2003). Factors influencing contractor performance: an international investigation. *Engineering, Construction and Architectural Management, 10*(5), 322–332.

Xu, Y., Chan, A. P. C., & Yeung, J. F. (2010a). Developing a fuzzy risk allocation model for PPP projects in China. *Journal of Construction Engineering and Management, 136*(8), 894–903.

Xu, Y., Yeung, J. F., Chan, A. P. C., Chan, D. W., Wang, S. Q., & Ke, Y. (2010b). Developing a risk assessment model for PPP projects in China—A fuzzy synthetic evaluation approach. *Automation in Construction, 19*(7), 929–943.

Yescombe, E. R. (2011). *Public-private partnerships principles of policy and finance.* Elsevier.

Yeung, J. F. Y. (2007). *Developing a partnering performance index for construction projects – A fuzzy set theory approach* (Doctoral thesis). Hong Kong: Hong Kong Polytechnic University.

Yeung, J. F., Chan, A. P., & Chan, D. W. (2009). Developing a performance index for relationship-based construction projects in Australia: Delphi study. *Journal of Management in Engineering, 25*(2), 59–68.

Yong, H. K. (Ed.). (2010). *Public-private partnerships policy and practice: A reference guide.* Commonwealth Secretariat.

Yuan, J., Zeng, A. Y., Skibniewski, M. J., & Li, Q. (2009). Selection of performance objectives and key performance Indicators in public-private partnerships projects to achieve value for money. *Construction Management and Economics, 27*(3), 253–270.

Yun, S., Jung, W., Han, S. H., & Park, H. (2015). Critical organizational success factors for public private partnership projects – A comparison of solicited and unsolicited proposals. *Journal of Civil Engineering and Management, 21*(2), 131–143.

Zawawi, M. I. Z., Kulatunga, U., & Thayaparan, M. (2016). Malaysian experience with public-private partnership (PPP): Managing unsolicited proposal. *Built Environment Project and Asset Management, 6*(5), 508–520.

Zhang, X. Q. (2004). Concessionaire selection: methods and criteria. *Journal of Construction Engineering and Management, 130*(2), 235–244.

Zhang, X. Q. (2005a). Criteria for selecting the private-sector partner in public–private partnerships. *Journal of Construction Engineering and Management, 131*(6), 631–644.

Zhang, X. Q. (2005b). Critical success factors for public-private partnerships in infrastructure development. *Journal of Construction Engineering and Management, 131*(1), 3–14.

Zhang, X. (2006a). Public clients' best value perspectives of public private partnerships in infrastructure development. *Journal of Construction Engineering and Management.* https://doi.org/10.1061/(ASCE)0733-9364(2006)132%3A2(107).

Zhang, X. (2006b). Factor Analysis of Public Clients' Best-Value Objective in Public–Privately Partnered Infrastructure Projects. *Journal of Construction Engineering and Management.* https://doi.org/10.1061/(ASCE)0733-9364(2006)132:9(956).

Zhang, X. Q., & Kumaraswamy, M. M. (2001). Hong Kong experience in managing BOT projects. *Journal of Construction Engineering and Management, 127*(2), 154–162.

Zhao, B. (2009). Land expropriation, protest, and impunity in rural China. *Focaal, 2009*(54).

Zouggari, M. (2003). Public-private partnership: Major hindrances to the private sector's participation in the financing and management of public infrastructures via delegated management. *International Journal of Water Resources Development, 19*(2), 123–129.

Index

A
Administrative, 42, 45–47, 94, 96
Africa, 18–23, 28, 31, 39, 43, 46, 110, 111, 115, 118, 121, 132, 137
Arbitration, 128, 135, 136, 141
Australia, 3, 4, 6, 9, 14–16, 31, 39, 40, 43, 80, 110, 113
Authorities, v, 2, 14, 15, 18, 22, 23, 40, 58, 59, 66, 78, 111, 118, 119, 122, 124, 132–136

B
Best practice framework, vi, 130, 145–149
Best practices, v, 39, 51, 66, 74, 145–149
Bidders, 4, 6, 111, 122, 123
Budget, 38, 93, 94, 96, 97, 99, 100, 102
Build operate transfer (BOT), 3, 20, 24, 72
Build own operate, 3, 15, 132
Build transfer and operate (BTO), 3, 4
Business, 19, 20, 23, 26, 28, 30, 43, 63, 82, 136, 138

C
China, 3, 18, 31, 38, 40, 63, 64, 72, 73, 110, 113, 127, 129–136, 141
Civil group societies, 81, 137
Commitments, 6, 17, 18, 20, 23–25, 30, 52, 54, 59, 61, 66, 74, 77, 79, 80
Communication, 23, 58–60, 128, 131, 134, 135, 137, 139–141
Compensation, 54, 82, 103, 111, 130, 131, 133, 134
Competition, 23, 24, 30, 52, 62, 74, 77, 79–81, 85, 103, 114, 115, 117, 118, 121, 123
Completion, 3, 4, 22, 24, 62, 66, 72, 74, 77, 79–83, 86, 87
Concessions, 2–4, 6–8, 14, 16, 20–22, 24, 57, 65, 81, 135, 137
Conflict management, vi, 127–142
Conflict prevention, 127, 136–142
Conflict resolution methods, 136
Consortium, 2–4, 6–8, 20–23, 28, 29, 45, 57, 59, 60, 63, 65, 78
Construction, 3, 6, 8, 10, 16, 19, 21, 22, 24–28, 30, 43–45, 53, 57, 62, 72–75, 78–80, 82–84, 86, 87, 92, 93, 102, 128, 133, 134, 141
Construction stage, 4, 6, 84, 148
Contracting authorities, 4, 7, 66, 104, 110, 111, 115, 122, 123, 134, 136–138, 145
Corruption, 53, 62, 74, 75, 78–81, 86, 95, 109, 114, 117, 118
Costs, 3, 4, 7–9, 14, 20–27, 29, 30, 38, 39, 42, 44–47, 55, 56, 62, 63, 72–76, 78–84, 87, 92, 94, 96–102, 104, 105, 111, 114, 116, 117, 119, 120, 122–124, 128, 134, 135
Countries, v, 1, 3, 4, 6, 8, 9, 13–22, 26, 29, 31, 38–46, 51–57, 61–67, 72–74, 79–82, 84, 86, 87, 92, 102–105, 109–113, 115–118, 121–123, 129, 130, 132, 137–140
Critical, 21, 38, 41, 42, 46, 47, 54, 62–65, 67, 71, 72, 79, 82–86, 95–98, 102–105, 114–116, 119–122, 124, 130, 132, 134, 136–142

R. Osei-Kyei, A. P. C. Chan, *International Best Practices of Public-Private Partnership*, https://doi.org/10.1007/978-981-33-6268-0

Critical success factors (CSFs), vi, 52–67
Cross harbour tunnel (CHT), 19, 24–25, 31, 43

D
Data, 19, 37, 39–40, 51, 71, 91, 95–97, 105, 112–113, 129–130
Definitions, v, 1, 2, 9, 74, 78–79, 92, 93
Delivery, 1, 2, 8, 14, 15, 18, 23, 25–28, 30, 39, 42–47, 60, 74, 78, 80, 81, 86, 95, 118, 120, 123, 131, 137, 139, 141
Design build finance operate (DBFO), 3, 16
Designs, 3, 4, 16, 20, 27, 44, 60, 63, 65, 73, 74, 76, 78–80, 84, 86, 128
Developed economies, v, vi, 15–17, 31, 37, 39, 41, 43, 47, 51, 56, 67, 71, 73
Developing economies, v, vi, 17–19, 31, 38, 43
Development, v, 2, 4, 8, 9, 14–18, 20–25, 31, 39, 40, 42–45, 47, 53–56, 59, 60, 63, 65, 66, 72, 73, 84, 93, 94, 96–101, 103–105, 111, 114, 119, 122, 123, 134, 138
Disputes, 53, 54, 58, 65, 78, 94, 96–101, 103–105, 128, 130, 132, 134–139, 141
Dysfunctional conflict, 128

E
Economics, 3, 14, 18, 20, 21, 24, 25, 38, 42–45, 47, 51, 53–55, 58, 60, 62–64, 67, 73, 74, 82, 83, 86, 94, 96, 97, 99, 100, 103, 104, 110, 116, 138
Efficiency Unit, 2, 7, 39, 40, 61, 66
Evolution, 13, 14, 30, 31, 51
Expression of Interest (EOI), 6

F
Facilities, 2–4, 6–9, 14, 15, 20–22, 24–27, 30, 31, 43–46, 63, 78, 79, 81–85, 94–96, 103, 104, 114–116, 118, 122, 133, 139
Factor analysis, 96, 97, 102, 103, 139
Feasibility, 4, 6, 9, 23, 58, 63, 102–104, 115, 133, 134
Financial, 4, 6, 8, 9, 15, 19–23, 26–28, 30, 31, 38, 39, 42, 45, 53–58, 60–62, 65–67, 72, 73, 78, 81–83, 93, 110, 114, 115, 117, 120, 124, 132
Financial close, 6, 8, 18, 19, 103
Financiers, v, 8–9, 22
Functional conflict, 128
Funds, 8, 18, 20–23, 27, 28, 30, 37, 42, 55, 57, 66, 78, 84, 115
Fuzzy, 98, 99, 105

G
Ghana, 9, 19, 26–31, 37–47, 51, 59–67, 71, 74–86, 91, 104, 105, 113, 127–136, 139–141
Ghana National Housing Project (GNHP), 19, 26–31, 63, 81, 85, 128, 132
Global implementation, 13–31
Government of Ghana, 2, 42, 132
Governments, v, 2–4, 6, 8–10, 13–30, 37–47, 52–53, 55–63, 65–67, 72, 73, 78, 81–83, 87, 103, 110, 111, 114–121, 123, 124, 132–134, 136, 138
Groupings, 45, 47, 64, 95–98, 105, 139–141

H
Hong Kong, 2, 3, 9, 14, 19, 24–26, 31, 37, 39–47, 51, 59–67, 71, 73–86, 103, 113, 128
Hospitals, 7, 15, 22–24, 31
Housing, 26–30, 40, 92

I
Incomes, 20, 24, 27, 29, 55, 94
Index, 55, 82, 91–105, 119
Indicators, 55, 60, 62, 65, 82, 83, 92, 93, 101
Inflation, 54, 55, 62, 72, 74, 76, 78, 80–82, 86
Infrastructure, v, 2–4, 6, 7, 14–20, 23, 25, 28, 37–39, 41–47, 63, 72, 93, 110, 114–116, 118, 122
Innovations, 25, 26, 38, 42–45, 47, 56, 60, 62, 65–67, 93, 94, 96, 102, 115, 119, 120, 124, 138
Institutional literatures, 116
Intellectual property, 56, 114, 117, 119, 120, 122, 124
International, v, vi, 1, 13, 21, 23, 25, 27, 28, 30, 38, 39, 42–44, 56, 57, 63, 66, 74, 82, 87, 93, 112, 124, 130, 133
International case studies, 19–31
Investments, 4, 6, 14, 16–19, 21, 22, 24, 25, 38–40, 42, 43, 52, 55, 56, 62, 63, 66, 72, 78, 82, 84, 102, 103, 116, 130, 132
Investors, v, 2–4, 6, 8, 14, 15, 27, 28, 30, 42–44, 46, 52–56, 58, 62, 63, 66, 78, 81, 83, 84, 86, 87, 103, 110, 114–118, 120–122, 124, 133, 134, 136–139

J
Joint ventures, 25, 28, 61, 134, 135
Jurisdictions, 6, 9, 14, 37, 39, 41, 45–47, 52, 55, 57, 59, 66, 67, 74, 129, 130

L
Land acquisition, 72–74, 76, 78–80, 83, 84, 87
Legal framework, 64, 65
Litigations, 79, 94, 96, 97, 99, 100, 103, 104, 136
Local, v, 2, 6, 9, 15, 16, 18, 19, 21–23, 26–30, 42, 43, 45, 47, 54, 56, 57, 62, 66, 78, 94, 96–101, 103–105, 116, 118, 133–139, 141

M
Management, vi, 6, 8, 10, 21, 23, 26, 30, 42, 44, 52, 53, 57, 74, 78, 79, 87, 93, 103, 104, 109–124, 127, 128, 134, 135, 141
Markets, 13–16, 18, 19, 28, 29, 31, 42, 43, 54, 55, 57, 60–62, 66, 73–75, 78–80, 84, 115, 130, 132, 135
Means, 38, 41–46, 59, 60, 62, 64–66, 72, 73, 86, 95, 96, 98, 99, 105, 114–122, 130–132, 134–139
Modalities, 10, 65
Models, vi, 1, 3–4, 14, 15, 20, 30, 55, 91–105
Motivations, v, vi, 37–39, 41–46, 109, 110, 114–118, 124

N
Negotiations, 6, 9, 28, 53, 58, 78, 79, 128, 132, 133, 135, 136, 138, 140, 141
Nigeria, 19–22, 43, 62, 73, 81, 82, 113, 137

O
Operation and maintenance, 3, 8
Operation and transfer stage, 4, 6–7, 10
Operations, 3, 6, 15, 21, 23–26, 56, 72, 74, 76, 78–80, 84, 86, 94–97, 99, 100, 103, 104, 138

P
Parliament, 30, 132, 135
Parties, 1, 2, 6–10, 21–23, 25–30, 46, 52, 53, 55, 58–60, 63, 65, 73, 74, 77–82, 85, 94, 95, 103, 104, 121, 122, 130–132, 136–142
Partnerships, 2, 6, 13–31, 43, 51–67, 71–87, 91–105, 109–124, 137, 141, 145–149
Performances, vi, 7, 19–23, 26, 38, 58, 91–94, 96, 97, 99–101, 103, 104, 128, 133, 137, 138, 140
Policies, v, 3, 4, 13–18, 26, 29, 38, 39, 43, 44, 52, 54–56, 60, 61, 64, 65, 73, 79, 82, 83, 91, 103, 114, 115, 117, 119–122, 124, 133, 139, 148
Policymakers, 55, 119, 120, 137
Political, 15, 21, 28, 29, 31, 37, 38, 52–53, 60, 61, 63–65, 72–76, 78–82, 84–87, 95, 96, 103, 121, 122, 130–133
PPP definitions, 2
PPP international case studies, 13
PPP models, 1, 3, 10, 20
PPP process, 5, 6, 58, 60–62, 65, 102, 122, 130, 132, 145–149
PPP stakeholders, v, 7, 9, 23, 53, 54, 71, 91, 93, 105, 121, 127, 128, 137, 142
Practitioners, vi, 1, 37, 39, 40, 42, 52, 56, 62, 66, 72, 91, 94, 98, 101, 103–105, 110, 112, 113, 119, 120, 127, 129, 130, 137, 139
Private, v, 2–4, 6, 8, 9, 14, 15, 17–25, 28–30, 40–46, 52–60, 62, 63, 65, 66, 73, 74, 76, 78–84, 93–95, 102–104, 110, 111, 114–118, 120–122, 124, 129, 132–134, 136, 138–140, 145, 148
Private developers, v, 3, 4, 26, 27, 29, 30, 54, 56, 102, 115
Private sector innovations, 114, 117, 120, 124
Process, v, 1, 4–7, 10, 19, 21–23, 28–30, 52–54, 56, 58, 60, 61, 65, 71, 72, 78, 79, 81, 85, 93, 110, 111, 114, 116, 117, 119, 121, 122, 124, 128, 132, 133, 140, 141
Procurements, v, 4, 10, 15, 16, 18, 20, 23, 28–30, 39, 40, 44–46, 52, 53, 56, 58, 62, 63, 71, 74, 78, 79, 81, 84, 93, 102, 109–111, 122–124
Procurement stage, 4, 6, 147
Profits, 15, 54, 83, 92, 94, 97–103, 105, 138
Project inception, 4, 10, 146
Project parties, 1–3, 7–9, 19, 23, 30, 58, 62, 72, 74, 77, 79, 80, 82, 83, 91, 104, 127, 128, 133, 135, 136, 138–141
Projects, v, vi, 1–10, 14–31, 38–40, 42–47, 51–67, 71–87, 91–105, 109–124, 127–142, 145
Project success index (PSI), 95–105
Proposals, vi, 4, 6, 22, 110, 111, 115, 116, 119–124

Public, v, 2–4, 6–10, 14–18, 21–23, 25, 27, 29–31, 38–47, 52–55, 58–67, 72–75, 78–85, 93–96, 102–104, 109–111, 114, 115, 117, 118, 120–122, 124, 129, 132–141
Public consultations, 119–121, 124
Public-private partnership (PPP), v, vi, 1–10, 13–26, 28–31, 37–47, 51–67, 71–87, 91–105, 109, 110, 112–121, 124, 127–142, 145–148

Q
Qualities, 6, 42–45, 47, 54, 74, 76, 79, 80, 92, 93, 95–97, 99, 100, 102–104
Questionnaires, 37–40, 51, 71–73, 91, 93, 110, 112, 124, 127, 129

R
Rankings, 41, 42, 45–47, 59–66, 74–86, 95, 96, 114, 119, 130, 133–136
Reasons for PPP, 38, 39, 42, 43
Regulations, 4, 22, 53, 56, 57, 64, 65, 74, 75, 78, 80, 134
Rehabilitate operate transfer (ROT), 20
Relationship, 2, 23, 58, 59, 63, 64, 67, 72–74, 79, 86, 94, 96, 97, 104, 128, 136, 139, 140
Resolution, vi, 54, 127–129, 135–136, 141
Respondents, 39–47, 51, 59, 61–64, 66, 67, 71, 82, 83, 86, 91, 99, 112–114, 116, 118, 119, 122–124, 129, 130, 132, 133, 135, 136
Revenue, 22, 24, 26, 72, 78, 79, 132, 133
Risk allocation, 15, 38, 58, 60, 63, 65, 72, 73, 84, 104, 120, 131, 134
Risk assessment, vi, 71–87, 104, 139, 140
Risk factors, 71–87, 129
Risk impact, 74, 80, 85, 86
Risk management, 46, 71–74, 95–97, 99, 100, 103, 104, 129, 134, 141
Risks, 2, 3, 6, 9, 24, 26, 31, 38, 42, 45–47, 52, 57, 58, 71–74, 77–81, 83–86, 94, 95, 102, 104, 115, 116, 119, 120, 124, 128, 129, 131, 132, 134, 137, 140

S
Services, 2, 3, 6, 7, 9, 15, 16, 22, 23, 25, 27, 40, 42–45, 47, 58, 60, 73, 78, 84, 93–97, 99, 100, 103, 104, 120, 131, 137–139, 141
Significance, 37, 41, 59, 61, 66, 73, 97, 116, 118, 119, 130
Similarities, v, 37, 39, 45–47, 64–66
Solicited, 109, 110, 116
Stakeholders, v, 7–9, 19, 21, 23, 28, 30, 53, 54, 57, 60, 71, 72, 79, 91–94, 103, 105, 119–121, 124, 127, 128, 130, 135, 137, 140–142
Success, vi, 13, 16, 20–26, 38, 51–67, 72, 83, 84, 86, 87, 91–105
Success criteria, 92–98, 102–105
Success model, 91–105
Surveys, 37–41, 51, 71–73, 91, 93, 96–98, 110, 112, 124, 127, 129
Sustainability, 44
Swiss challenge, 111, 123, 124
Systems, 14, 18, 20, 25, 42, 43, 46, 52, 54, 56, 72, 78, 80, 81, 84, 110, 111, 118, 123, 124, 128, 132, 133, 137, 140, 141

T
Tariffs, 73–75, 78, 80, 95, 116, 131, 134, 135, 137, 138, 140, 141
Technologies, 2, 22, 25, 39, 42–45, 47, 56–57, 60, 62, 65–67, 74, 75, 78, 80, 84, 93, 94, 96, 120, 138
Tendering, 22, 23, 52, 53, 60, 61, 65, 79, 85, 102, 110, 111, 116, 119, 121–124, 140
Tendering methods, 111, 122–124
Traditional, 42–46, 71, 82, 84, 85, 92, 102, 109
Transparency, 15, 53, 62, 64, 84, 85, 95, 103, 104, 118, 120, 121, 123, 135, 136, 139–141
Transport, 3, 16, 20, 21, 128

U
United Kingdom, 2, 110
United Nations, 53
United States, 110
Unsolicited, 109–124
Unsolicited PPPs, vi, 109–111, 114–124
Unsolicited proposals, 4, 110, 114–124
Urbanization, 44
User fees, 6, 9, 21, 23, 60, 63, 65, 78, 103

V
Value for money, v, 15, 23, 29, 30, 46, 72, 93, 94, 109, 110, 115, 119, 120, 124

W
World Bank, 14, 17–19, 40, 102, 103, 115, 118, 120, 130, 133